Digital Techn and Systems

Digital Techniques and Systems

D C Green
M Tech, C Eng, MIERE
Senior Lecturer in Telecommunication Engineering
Willesden College of Technology

Second edition

with self-test questions

Longman Scientific & Technical

Longman Scientific & Technical
an imprint of
Longman Group Limited
Longman House, Burnt Mill, Harlow
Essex CM20 2JE, England
Associated companies throughout the world

First printed in Great Britain by Pitman Publishing Limited 1980
Second edition 1982
Reprinted 1983
Reprinted with addition of self-test questions 1985
Reprinted by Longman Scientific & Technical 1986

ISBN 0-582-98836-5

Printed in Great Britain at The Bath Press, Avon

Contents

Preface

to the Second Edition

The second edition of *Digital Techniques and Systems* incorporates all of the changes in the content of the unit that have been introduced by the Business and Technician Education Council.

The original Digital Techniques III half-unit has been replaced by two half-units, Digital Techniques III A and B. Full coverage of both of the new half-units has necessitated a number of additions to the book plus a new section (in Chapter 6) on semiconductor memories.

The BTEC Transmission Systems III half-unit has in the main been merely re-arranged but a completely new section on Optical Fibres has been added. The coverage of this topic has meant the inclusion of an extra chapter (Chapter 11).

The opportunity has also been taken to correct a number of errors and omissions that had come to light.

D.C.G.

Note to 1985 reprint
A 24-page self-testing section has been added for this reprint. Three kinds of exercise are provided: true/false, multiple choice, and short conventional. Answers for the first two types are given at the end of the section. Its five parts correspond with the digital techniques chapters of the main text (Chapters 2–6).

Preface

This book provides a fairly comprehensive coverage of the basic techniques used in modern digital circuitry and of the elementary principles of data communication. The treatment is such that the book constitutes a suitable first course in the subject for technicians.

The Business and Technician Education Council (BTEC) scheme for the education of electronic and telecommunication technicians introduces the student to the concepts of digital circuits and data transmission at the third level in two separate half-units. This book has been written to provide a complete coverage of the Digital Techniques III and Transmission Systems III half-units.

Chapter 1 provides a basic introduction to some of the many and varied uses of digital techniques in modern technology and is intended to give the newcomer to the field some idea of the many possibilities associated with the use of modern digital circuitry. Chapters 2 and 3 cover the subject of electronic gates of all kinds; Chapter 2 discusses the various kinds of gate, e.g. AND and OR, and some of the ways that they can be interconnected to perform various logical functions. Chapter 3 introduces the various logic families with particular emphasis being placed upon t.t.l. and c.m.o.s. logic. The next three chapters deal with, respectively, flip-flops, counters, and ferrite core stores.

The remainder of the book is concerned with the transmission of data over telephone lines. Chapter 7 discusses the elementary principles of d.c. pulse transmission over lines and goes on to consider why modems are generally used for the majority of data links. Chapter 8 deals with the various forms of modulation that are used with data systems and then Chapter 9 introduces the reader to some typical data circuits. Lastly, Chapter 10 considers pulse code modulation which is increasingly used in modern telecommunication networks.

The book has been written on the assumption that the reader will possess a knowledge of electronics and telecommunication transmission techniques equivalent to that reached by the TEC level II units, Electronics II and Transmission Systems II. The reader is also assumed to have studied, or be concurrently studying, the level III unit, Electronics III, since a knowledge of integrated circuits is throughout taken for granted.

Acknowledgement is due to the Technician Education

The following abbreviations for other titles in this series are used in the text:

TSII: Transmission Systems II
RSII: Radio Systems II
RSIII: Radio Systems III
EII: Electronics II
EIII: Electronics III

Council for their permission to use the content of their units which are printed at the end of this book. The Council reserve the right to amend the content of their units at any time.

Some worked examples are provided in the text to illustrate the principles that have been discussed and each chapter concludes with a number of short exercises and longer exercises. A number of these exercises have been taken from past City and Guilds examination papers and grateful acknowledgement of permission to do so is made to the Institute. Answers to the numerical problems will be found at the end of the book; these answers are the sole responsibility of the author and are not necessarily endorsed by the Institute. Finally, a number of multiple choice questions are provided at the end of the book.

D. C. G.

1 Digital Systems

Most present-day electronic and telecommunication equipment is still analogue in nature. This means that the signals to be handled, processed or transmitted are represented by voltages whose amplitude and/or frequency vary continuously with time; thus, in a telephone system, the transmitted signals are replicas of the speech waveforms. Many examples of analogue equipment are well known; for example, the radio and television receivers to be found in the majority of homes.

Digital signals are not continuous in nature but consist of discrete pulses of voltage or current which represent the information to be processed. Digital voltages can vary only in discrete steps; normally only two voltage levels are used—one of which is zero—so that two-state devices can be employed. A two-state device is one which has only two stable states; so that it is either ON or it is OFF. Examples of two-state devices are: a lamp which is either glowing visibly or it is not; a buzzer which is either producing an audible sound or not; or an electrical switch which either completes an electrical circuit or breaks it. Further examples of two-state devices that are used extensively in electronic circuitry are the semiconductor diode and the transistor—bipolar and field effect [EII and EIII].

The advantages to be gained from the use of digital techniques instead of analogue methods arise largely from the use of just the two voltage levels. Digital circuitry, mainly integrated in modern systems, operates by switching transistors ON and OFF and does not need to produce or to detect precise values of voltage and/or current at particular points in an equipment or system. Because of this it is easier and cheaper to mass-produce digital circuitry. Also, the binary nature of the signals makes it much easier to consistently obtain a required operating performance from a large number of circuits. Digital circuits are generally more reliable than analogue circuits because faults will not often occur through variations in performance caused by changing values of components, misaligned coils, and so on. Again, the effects of noise and interference are very much reduced in a digital system since the digital pulses can always be regenerated and made like new whenever their waveshape is becoming distorted to the point where errors are likely. This is not possible in an analogue system where the effect of unwanted noise and interference signals is to permanently degrade the signal.

There are two main reasons why the application of digital techniques to both electronics and telecommunications has been fairly limited in scope until recent years. First, digital circuitry was, in the main, not economic until integrated circuits became freely available, and, secondly, the transmission of digital signals requires the provision of circuits with a very wide bandwidth. Some digital circuits and equipments have, of course, been available since pre-integrated circuit days but their scope and application were very limited.

The Digital Computer and the Microprocessor

Nowadays the digital computer is an integral part of the day-to-day operation of many firms and organizations, ranging from Government departments, commercial concerns such as banks and insurance companies, to industrial firms in all branches of engineering and science. Computers are employed for the calculation of wages and salaries, taxes, pensions, bills and accounts; for the storage of medical, scientific and engineering data; and for the rapid booking of aircraft seats, theatre tickets and foreign holidays. Computers are also used to carry out complex scientific and engineering calculations, to control engineering processes in factories, to control the operation of telephone exchanges and military equipment and weapons, to control the distribution networks for gas, water and electricity, and for many other purposes.

A digital computer is able to store large quantities of data in its memory which can be made available as and when required. The task to be performed by the computer is detailed to the computer by means of a set of instructions known as the *programme*. The programme is fed into the computer and stored in another part of its memory.

The basic block diagram of a digital computer is shown in Fig. 1.1. The instructions contained in the programme are taken sequentially (one after the other) from the memory under the direction of the control unit. Each instruction causes the arithmetic unit to perform arithmetic and logic operations on the data, also taken from the memory. The results of the calculations can be stored in the memory or they can be held temporarily in a part of the arithmetic unit known as an accumulator. When a calculation has been completed, the control unit will transfer the results to an output device, which will (probably) produce the results in printed form; alternatively the results may be transmitted over a *data link* to a distant point where they are needed.

The input devices used to feed information into a computer are usually some kind of teleprinter, or a paper tape or card reader, or a magnetic tape reader. Binary information can be stored on a paper tape by punching holes in the tape; a hole in the paper tape represents binary 1 and the absence of a hole

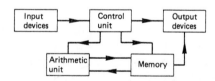

Fig. 1.1 Digital computer

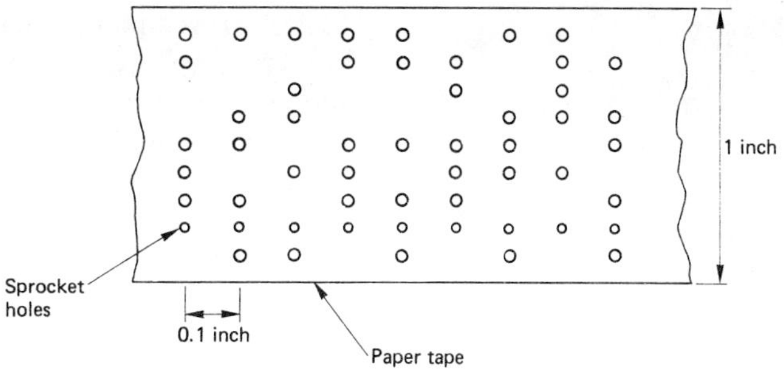

Fig. 1.2 Punched tape

1 inch

0.1 inch

Sprocket holes

Paper tape

indicates binary 0. The number of holes, and absences of holes, needed to represent a character depends upon the code that is used but Fig. 1.2 shows how a punched tape will look. The top binary digit or *bit* is a parity bit which is used to detect any errors that may exist. The remaining seven bits represent the character to be signalled to the computer. The smaller hole, labelled sprocket hole, that exists in every row is provided to engage the tape with the mechanism of the tape reader.

Teleprinters and other similar printing apparatus, and paper tape/card punchers, can be used as output devices but very much faster in operation are equipments known as line printers. Visual display units (VDUs) are also employed and these basically consist of a cathode ray tube upon the screen of which data can be displayed.

Minicomputers

A minicomputer has a smaller storage capability than a mainframe computer but the capacity provided is adequate for very many applications. A minicomputer is small enough physically to be brought into an office, often on a trolley, when it is needed, plugged into the electric mains supply, and then used. Minicomputers are also used as relatively inexpensive means of controlling industrial systems and processes.

Microprocessors

The term microprocessor is generally applied to a form of minicomputer that is able to control the operation of a wide variety of equipments. A microprocessor occupies very little space since much of its circuitry is contained within an integrated circuit. As a result, a microprocessor can be built into the equipment whose operation it is to control. Fig. 1.3 shows the basic block diagram of a microprocessor; the chip contains various registers, an arithmetic unit and control circuitry very similar to a main computer. Often the memory is also integrated.

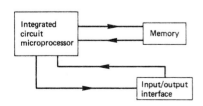

Fig. 1.3 Microprocessor

As an example, microprocessor control of modern radio receivers and systems is increasingly used in the latest equipments. A microprocessor can be programmed to control the tuning, the gain and the selectivity of a receiver as the receiving conditions alter. Remote h.f. radio stations can be distantly controlled by means of a microprocessor; the functions controlled being the selection of the frequency to be transmitted and the frequency to be received as determined by a pre-set schedule. Also, the performance of the station can be monitored and any faults or degradation of service detected and recorded.

Digital Equipment and Systems

Hand-calculators are nowadays in common use and provide another example of the use of digital circuitry. All calculators are able to carry out the basic mathematical procedures while many are provided with several more advanced mathematical facilities. Some models are programmable. The circuitry contained within a calculator is complex and these devices have only become practical since the advent of integrated circuits.

Many cash registers and weighing scales used in the shops are electronic and these provide a readout in digital form of the total money to be paid and, once the money offered has been entered by the operator, of the change to be given, as well as a printout of the purchases. The cash register also keeps a record of the cash input and in some cases this is signalled to a central point where a complete record for the shop or department can be maintained. Often a microprocessor is also involved so that records are automatically kept of the items sold, and the stock left on the shelves and in the store. This enables the management to know at all times the stocks held of all items so that re-stocking can be carried out in ample time. It is possible to programme the microprocessor to order from the warehouse those items whose stocks are falling below an appointed figure.

In engineering, digital readouts of data are often more convenient and accurate than analogue readings. Digital voltmeters and frequency meters or counters are particularly suited to measurement applications where a large number of repetitive readings are to be made. The advantage of digital meters is most noticeable when relatively unskilled personnel are employed to carry out the tests. With a digital instrument the operator can read at a glance the value of the displayed parameter, but very often an analogue reading requires care if reasonable accuracy is to be obtained. This point is illustrated by Fig. 1.4. When the pointer (Fig. 1.4a) is in between two scale markings, some doubt exists as to the value indicated. No

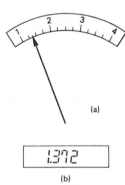

(a)

(b)

Fig. 1.4 Indication of measured quantity by (a) analogue, (b) digital method

such doubt is present with the digital instrument; its indicated value (Fig. 1.4b) is easy to read.

Data-loggers are digital circuits that convert the analogue output from a transducer (a resistance strain gauge, a thermocouple, a potentiometer for example) into a digital form so that the measured parameter can be recorded on a paper tape or some other means.

Another application of computers and digital techniques that is gaining in importance in the modern world is in the field of transport. The movement of vehicles in a large transport system can be controlled and monitored by a computer. British Rail, for example, have introduced a computerized system for the optimised control of its freight traffic. Each truck has its movements continuously monitored and the computer works out and augments the best way of moving the trucks around the network in order to carry the maximum amount of freight in the most economic manner. In many large cities, computers control the traffic lights that direct the flow of traffic across road junctions. The computer continuously monitors the number of cars passing and waiting to pass the various junctions and varies the frequency of the traffic light operations to optimise the flow of traffic.

Telephone Transmission Systems

Telecommunication systems have traditionally been analogue in their nature, with the exception of Morse code telegraphy. Speech signals are transmitted over purely analogue circuits routed over a combination of physical pairs in telephone cables and frequency division multiplex channels over line and/or radio links. Digital transmission using *pulse code modulation* is increasingly used and much of the junction network in the United Kingdom now uses this technique. The future development of the trunk network of the U.K. is destined to use digital techniques and British Telecom have announced the introduction of SYSTEM X. System X is to be an integrated telephone network in which junction and trunk transmission, signalling and exchange switching are all to be achieved using digital methods under the control of digital computers. One advantage of digital working is that it can be used for all kinds of signal, be it speech, music, television, telegraphy or data. Other advantages are discussed in Chapter 10.

The distribution of the frequency-modulated v.h.f. sound broadcast signals and the audio signals of television broadcasts from studios to transmitters is carried out digitally using pulse code modulation. Teletext services are now transmitted by both the B.B.C. and the I.B.A.—using the names CEEFAX and ORACLE respectively—to provide information to the home. The data is transmitted digitally using some of the lines in each field of the television signal which are not modulated by the video signal. In the television receiver a digital decoder is provided to recover and display the incoming information on the television screen. A similar kind of information service, known as PRESTEL, has been introduced by British Telecom. The data is transmitted, again in digital form, over a telephone line to the home and, after decoding, is displayed by the television receiver.

Private mobile radio systems operating in the v.h.f. and the u.h.f. bands can also be controlled by a computer to ensure the optimum performance as the mobiles move around the service area. Such systems are used by organizations that employ a large number of mobiles, such as the Gas Board in the U.K.

2 Electronic Gates

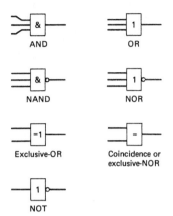

Fig. 2.1 Gate symbols

An electronic gate is a circuit that is able to operate on a number of input binary signals in order to perform a particular logical function. The logic gate is one of the basic building blocks from which many different kinds of logical system can be constructed. Electronic gates are readily available in integrated circuit form and the various logic families in common use will be discussed and their characteristics compared in the next chapter. In this chapter the emphasis will be on the various types of gates and the ways in which they can be interconnected to perform different logical functions. The types of gate to be considered are the AND, NOT, OR, NAND, NOR, exclusive-OR, and the exclusive-NOR or coincidence gate. The British Standard symbols for each of these gates are given in Fig. 2.1. Positive logic is assumed throughout this chapter; that is, logic 1 is represented by the more-positive voltage, and logic 0 by the less-positive voltage, of two possible values.

The AND Gate

The AND gate is a logic element having two or more input terminals and only one output terminal. The output logic state is 1 only when *all* of the inputs are at logic 1. If any one or more of the inputs is at logic 0, the output state will also be logical 0. Using Boolean algebra the output F of an AND gate with three inputs A, B, and C can be written down as

$$F = A B C \qquad (2.1)$$

since in Boolean algebra the symbol for the AND logical function is the (omitted) dot (.).

The operation of any logical element can be described by means of a **truth table**; this is a table which shows the output state of the circuit for all the possible combinations of input states. The truth table of a 3-input AND gate is given by Table 2.1. It is clear from the table that the output is 1 only when A AND B AND C is 1.

The AND gate can be used to *enable* and *inhibit* a digital signal. Since the output of a 2-input AND gate will be 1 only if both its inputs A and B are 1, a control signal applied to, say, input A can control the passage of a signal applied to input B. When input A is at the logic 0 level, it will stop, or **inhibit**, the signal at B from passing through the gate. When input A is at logical 1, it will allow, or **enable**, the signal applied to B to pass to the output. Very often the control signal is a regularly

Table 2.1 AND gate

A	0	1	0	0	1	1	0	1
B	0	0	1	0	1	0	1	1
C	0	0	0	1	0	1	1	1
F	0	0	0	0	0	0	0	1

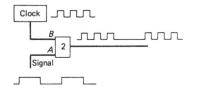

Fig. 2.2 Use of an AND gate to control the passage of a clock waveform

occurring pulse waveform derived from a circuit known as the *clock*. An example of this technique is shown in Fig. 2.2.

The output of an AND gate is always either logical 1 or 0 and cannot be at any in-between voltage. An **analogue gate** is one which acts as an electronic switch to allow an input analogue signal to pass when a control voltage is at logical 1 and to close the transmission path when the control voltage is at logical 0. Such a gate, used for example in the production of a pulse amplitude modulated signal, can be made with discrete components as shown by the circuit given in Fig. 2.3, or is available in integrated circuit form, e.g. the c.m.o.s. 4016. In the circuit of Fig. 2.3 the input analogue signal is only present at the inverting terminal of the operational amplifier [EIII] when transistor T_1 has been turned OFF by a clock pulse.

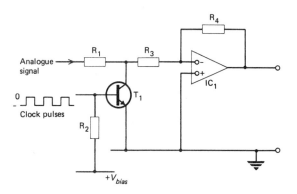

Fig. 2.3 An analogue gate

The OR Gate

An OR gate has two or more input terminals and a single output terminal which will be at logic 1 whenever any one or more of its input terminals is at logical 1. The Boolean expression for the output of a 3-input OR gate is given by equation (2.2), the OR function being represented by the symbol +.

$$F = A + B + C \qquad (2.2)$$

The truth table of a 3-input OR gate is given by Table 2.2.

Table 2.2 OR gate

A	0	1	0	0	1	1	0	1
B	0	0	1	0	1	0	1	1
C	0	0	0	1	0	1	1	1
F	0	1	1	1	1	1	1	1

EXAMPLE 2.1

Write down the Boolean expression representing the circuit shown in Fig. 2.4. With the aid of a truth table determine the necessary input state for the output to be at logical 1. Deduce a simpler arrangement that will perform the same logical function.

Solution

$$F = (A\,B\,C)(A + B + C)$$

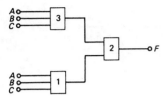

Fig. 2.4

The truth table is

A	B	C	A B C	A + B + C	F
0	0	0	0	0	0
1	0	0	0	1	0
0	1	0	0	1	0
0	0	1	0	1	0
1	1	0	0	1	0
1	0	1	0	1	0
0	1	1	0	1	0
1	1	1	1	1	1

It is clear from the truth table that the output of the circuit is at logic 1 only when A and B and C are all at 1. Thus the logic function of the circuit can more easily be performed by a 3-input AND gate alone. This is the first indication that very often a given logical function can be simplified and produced by some simpler arrangement.

Boolean equations can be simplified by either algebraic or mapping methods. Algebraic simplification of logic functions is facilitated by the use of the **logic rules** which follow.

1 $A + \bar{A} = 1$
2 $A + A = A$
3 $A A = A$
4 $A \bar{A} = 0$
5. $A(B + C) = A B + A C$
6 $A + 0 = A$
7 $A + 1 = 1$
8 $A 1 = A$
9 $A 0 = 0$
10 $A B = B A$
11 $A + B = B + A$
12 $B(A + \bar{A}) = B$
13 $A + A B = A$
14 $A(A + B) = A$
15 $A + \bar{A} B = A + B$
16 $A(\bar{A} + B) = A B$
17 $\overline{A + B} = \bar{A} \bar{B}$
18 $\overline{A B} = \bar{A} + \bar{B}$

All of these rules can easily be confirmed by the use of a truth table, as will now be shown in the cases of rules 17 and 18. These two rules are known as **De Morgan's rules** and are particularly useful. The truth table for rule 17 is given in Table 2.3. Obviously the fourth and the last columns are identical so that, for all values of A and B,

$$\overline{A + B} = \bar{A} \bar{B}$$

Table 2.3 De Morgan's first rule

A	B	A + B	$\overline{A+B}$	$\bar{A}$	$\bar{B}$	$\bar{A}\bar{B}$
0	0	0	1	1	1	1
1	0	1	0	0	1	0
0	1	1	0	1	0	0
1	1	1	0	0	0	0

Table 2.4 De Morgan's second rule

A	B	A B	$\overline{AB}$	$\bar{A}$	$\bar{B}$	$\bar{A}+\bar{B}$
0	0	0	1	1	1	1
1	0	0	1	0	1	1
0	1	0	1	1	0	1
1	1	1	0	0	0	0

Table 2.4 gives the truth table for the De Morgan's second rule, i.e. rule 18. Again, the fourth and last columns are identical and hence the rule is confirmed.

It is necessary to be able to express the output of a logical network in terms of Boolean algebra and also, given the Boolean expression for a required output, to be able to design the logic circuitry needed to produce this output. When designing a logic network, the required function is generally first simplified, using the rules 1–18, in an effort to minimize the number of gates required, although it must be noted that the simplest Boolean expression does not necessarily give the minimum number of gates.

EXAMPLE 2.2

Draw the logic circuit that will implement the Boolean expression $F = (A + B)(B + C)$. Also, simplify the expression and draw the circuit that will implement the simplified equation.

Solution

$$F = (A + B)(B + C)$$
$$= A B + A C + B B + B C$$
$$= A B + A C + B + B C \quad \text{rule 3}$$
$$= A B + A C + B(1 + C)$$
$$= A(B + C) + B \quad \text{rule 7}$$
$$= A C + B(1 + A)$$
$$= A C + B \quad \text{rule 7}$$

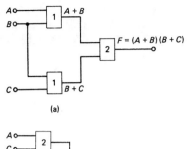

(a)

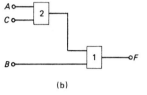

(b)

Fig. 2.5

Figs. 2.5*a* and *b* show, respectively, the logical circuitry required to implement the equation $F = (A + B)(B + C)$ and its simplified version $F = A C + B$. Clearly the saving in the number of gates needed is small, the number being reduced from three to two. If integrated circuit gates (see Chapter 3) are used, there might be no saving at all since a single integrated circuit package will contain four 2-input AND or OR gates.

The NOT Gate

The NOT gate has just one input and a single output terminal and it is used in logical circuitry as an **invertor**, i.e. the NOT logical function inverts its input. The Boolean expression describing a NOT function is the bar over the input signal as shown in equation (2.3). For input A

$$F = \bar{A} \tag{2.3}$$

EXAMPLE 2.3

Simplify the Boolean equation

$$F = \overline{\bar{A}(B + \bar{C})}(A + \bar{B} + C)(\overline{\bar{A}\,\bar{B}\,\bar{C}})$$

and draw the circuit which will implement the simplified equation.

Solution

$$
\begin{aligned}
F &= (A + \overline{\bar{B} + \bar{C}})(A + \bar{B} + C)(A + B + C) \\
&= (A + \bar{B}\,C)(A\,A + A\,B + A\,C + A\,\bar{B} + \bar{B}\,B + \\
&\quad \bar{B}\,C + A\,C + B\,C + C\,C) \\
&= (A + \bar{B}\,C)[A + A(B + C) + \bar{B}(A + C) + C(B + A) + C] \\
&= (A + \bar{B}\,C)[A(1 + B + C) + \bar{B}(A + C) + C(1 + B + A)] \\
&= (A + \bar{B}\,C)[A + \bar{B}\,A + \bar{B}\,C + C] \\
&= (A + \bar{B}\,C)[A(1 + \bar{B}) + C(1 + \bar{B})] \\
&= (A + \bar{B}\,C)(A + C) \\
&= A\,A + A\,C + A\,\bar{B}\,C + \bar{B}\,C \\
&= A(1 + C) + \bar{B}\,C(1 + A) \\
&= A + \bar{B}\,C
\end{aligned}
$$

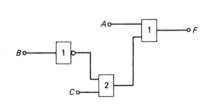

Fig. 2.6

Fig. 2.6 shows the required circuit.

The NAND Gate

The NAND gate performs the inverse of the AND logical function, so that its output is at logical 0 only when *all* of its inputs are at logical 1. The truth table of a 3-input NAND gate is given by Table 2.5. Clearly the output of a NAND gate is only at logical 0 when all three of its inputs are at logical 1. This action can be described by the Boolean expression

$$F = \overline{A\,B\,C} \tag{2.4}$$

Table 2.5   NAND gate

A	0	1	0	1	0	1	0	1
B	0	0	1	1	0	0	1	1
C	0	0	0	0	1	1	1	1
F	1	1	1	1	1	1	1	0

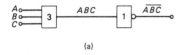

(a)

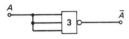

(b)

Fig. 2.7 The NAND function produced by (a) an AND gate followed by an inverter, (b) a NAND gate

Fig. 2.8 The NOT function produced by a NAND gate

Table 2.6 NOR gate

A	0	1	0	0	1	1	0	1
B	0	0	1	0	1	0	1	1
C	0	0	0	1	0	1	1	1
F	1	0	0	0	0	0	0	0

The NAND function can be produced by an AND gate followed by a NOT gate (Fig. 2.7a) but it is most often produced by a NAND gate (Fig. 2.7b). The NAND gate is readily available in integrated circuit form.

The NOT function can be obtained using a NAND gate by connecting its input terminals together (see Fig. 2.8). Equation (2.4) can be written, using De Morgan's theorem, as $F = \bar{A} + \bar{B} + \bar{C}$. This means that the NAND gate produces the same output signal as an OR gate fed with complemented (inverted) inputs.

The NOR Gate

The NOR gate performs the same logical function as an OR gate that is followed by an invertor. This means that its output is at logical 0 whenever any one or more of its inputs is at logical 1. The truth table of a 3-input NOR gate is given by Table 2.6 and the Boolean equation describing the function is given by expression (2.5).

$$F = \overline{A + B + C} \tag{2.5}$$

Rewriting equation (2.5) as $F = \bar{A}\,\bar{B}\,\bar{C}$ shows that the NOR gate can be regarded as an AND gate with complemented inputs. The NOR gate can also be used as an *inverter* or NOT gate by connecting its inputs together.

EXAMPLE 2.4

The waveforms shown in Fig. 2.9a and b are applied to (a) a 2-input AND gate, (b) a 2-input OR gate, (c) a 2-input NOR gate, and (d) a 2-input NAND gate. For each case draw the output waveform of the gate.

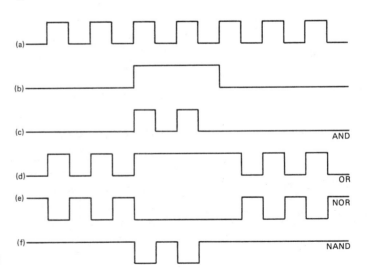

Fig. 2.9

Solution

The required waveforms are shown in Figs. 2.9*c*, *d*, *e*, and *f* respectively.

The Exclusive-OR Gate

The exclusive-OR gate has just two input terminals and one output terminal and it performs the logical function

$$F = A\bar{B} + \bar{A}B \qquad (2.6)$$

The truth table describing this function is given by Table 2.7. The output *F* of the gate is at logical 1 only when either one, but not both, of its inputs is also at logical 1. If both inputs are at logical 0 or at logical 1, the output of the gate will be at 0.

The exclusive-OR gate can be fabricated by suitably combining other types of gate and it can also be obtained in an integrated circuit package. Fig. 2.10 shows how an exclusive-OR gate can be made using a mixture of AND and OR gates.

The Coincidence Gate

The coincidence, or exclusive-NOR, gate is one which has two input terminals and one output terminal. It produces the logical 1 state at the output only when the two inputs are at the same logical state. The truth table of a coincidence gate is given by Table 2.8.

From the truth table it is apparent that the Boolean equation describing a coincidence gate is

$$F = \bar{A}\bar{B} + A B \qquad (2.7)$$

If equation (2.6) is inverted

$$\bar{F} = \overline{\bar{A}\bar{B} + A B}$$
$$= \overline{A\bar{B}} \cdot \overline{\bar{A}B}$$
$$= (A + B)(\bar{A} + \bar{B})$$
$$= A\bar{A} + A\bar{B} + B\bar{A} + B\bar{B}$$
$$= A\bar{B} + \bar{A}B$$

which is the equation for an exclusive-OR gate. Hence a coincidence gate performs the inverse function to an exclusive-OR gate.

The Use of NAND/NOR Gates to Generate AND/OR Functions

The majority of integrated circuit gates employed in modern equipment belong to one or other of two logic families, namely the **t.t.l.** and the **c.m.o.s.** families. In both of these families,

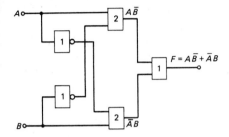

Fig. 2.10   Exclusive-OR gate

Table 2.7   Exclusive-OR gate

A	0	1	0	1
B	0	0	1	1
F	0	1	1	0

Table 2.8   Coincidence gate

A	0	1	0	1
B	0	0	1	1
F	1	0	0	1

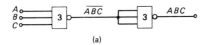

(a)

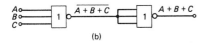

(b)

Fig. 2.11 Implementation of (a) the AND function using NAND gates, (b) the OR function using NOR gates

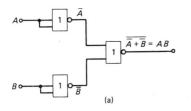

(a)

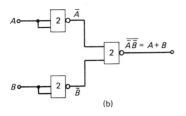

(b)

Fig. 2.12 Implementation of (a) the AND function using NOR gates, (b) the OR function using NAND gates

NAND and NOR gates are the most commonly used since their cost is less than that of the other types of gate which are available. Also, the NAND and NOR gates generally have a faster operating speed and a lower power dissipation. Very often therefore, a digital circuit is made up using *only* NAND and/or NOR gates.

It is easy to see that the AND function can be obtained by **cascading** two NAND gates as shown by Fig. 2.11a, the second gate being employed as an invertor. Similarly, the OR function is easily obtained by the cascade connection of two NOR gates (Fig. 2.11b).

Implementation of the AND function using NOR gates, and of the OR function using NAND gates, is not quite as easy but the necessary connections can readily be deduced by the use of De Morgan's rules. One rule is

$$\overline{A\,B} = \bar{A} + \bar{B}$$

hence

$$A\,B = \overline{\bar{A} + \bar{B}}$$

The right-hand expression is easily implemented using NOR gates as shown in Fig. 2.12a.

The other De Morgan rule is

$$\overline{A + B} = \bar{A}\,\bar{B}$$

hence

$$A + B = \overline{\bar{A}\,\bar{B}}$$

and this expression can be implemented using NAND gates as shown in Fig. 2.12b.

Clearly, more gates are needed to implement the AND/OR functions with NAND/NOR gates, but very often the apparent increase in the number of gates required is not as great as at first anticipated since consecutive stages of inversion need not be provided. This point is illustrated by the following example.

EXAMPLE 2.5

Implement the exclusive-OR function $F = A\,\bar{B} + \bar{A}\,B$ using (i) NAND gates only, (ii) NOR gates only.

Solution
The first step is to draw the logic diagram using AND, OR and NOT gates. When this has been done, replace each gate with its (i) NAND, (ii) NOR equivalent circuit. Finally, if possible simplify the resulting network by eliminating any redundant gates.

(i) Fig. 2.10 shows the exclusive-OR gate built with AND and OR gates. Replacing each gate with the equivalent NAND logic network gives the circuit of Fig. 2.13a. It can be seen that this network

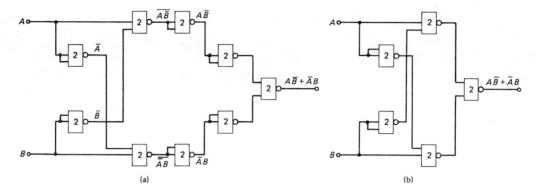

Fig. 2.13 Exclusive-OR function implemented using NAND gates

includes two sets of two NAND gates in cascade. These are redundant since $\bar{\bar{A}} = A$. Fig. 2.13b shows the simplified network which does not include the four redundant gates. It will be noticed that the number of NAND gates required (i.e. 5) to implement the exclusive-OR gate is the same as the number of AND and OR gates which are necessary.

(ii) Replacing each AND and each OR gate with the corresponding NOR gate version results in the circuit of Fig. 2.14a and this can be simplified by eliminating redundant gates to produce the final version given in Fig. 2.14b.

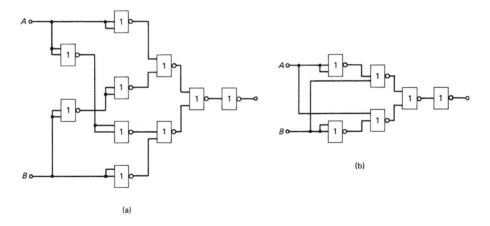

Fig. 2.14 Exclusive-OR function implemented using NOR gates

In general, Boolean equations in the *product-of-sums* form, e.g. $(A+B)(C+D)$, are best implemented using NOR gates, and *sum-of-product* equations, e.g. $A\,B+C\,D$, are more easily implemented using NAND gates.

The Karnaugh Map

The Karnaugh map provides a convenient method of simplifying Boolean equations in which the function to be simplified is displayed diagrammatically on a set of squares. Each square maps one term of the function. The number of squares is equal to 2^n, where n is the number of variables in the equation to be simplified. Thus, if the equation

$$F = A\,\bar{B} + \bar{A}\,B$$

is considered, then $n = 2$ and the number of squares needed is 4. The rows and columns of the map are labelled as shown so

that each square represents a different combination of the two variables. Thus the four squares represent, respectively,

$$A\,B \quad \bar{A}\,B \quad A\,\bar{B} \quad \bar{A}\,\bar{B}$$

The number 1 written in a square indicates the presence, in the function being mapped, of the term represented by that square. The number 0 in a square means that that particular term is not present in the function being mapped.

The mapping for the equation $F = A\,\bar{B} + \bar{A}\,B$ is

	A	Ā
B	0	1
B̄	1	0

To simplify an equation using the Karnaugh map, adjacent squares containing a 1 are looped together. When any two squares have been looped together, it means that the corresponding terms in the equation being mapped have been combined; and any terms of the form $A\,\bar{A}$ have been eliminated.

For example consider the equation $F = A\,B + \bar{A}\,B + \bar{A}\,\bar{B}$

	A	Ā
B	1	1
B̄	0	1

Looping adjacent squares in the maps as shown

simplifies the equation to

$$F = \bar{A}(B + \bar{B}) + B(A + \bar{A}) = \bar{A} + B$$

The Karnaugh map is easily extended for use with three variables A, B and C as shown by the following examples, or with four variables A, B, C and D. Squares must only be looped together in multiples of two, i.e. in twos, in fours, or in eights.

EXAMPLE 2.6

Use a Karnaugh map to simplify the equation

$$F = A\,C + \bar{A}\,B\,C + \bar{B}\,C$$

Solution
The Karnaugh map of the equation is

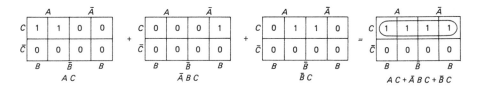

Normally of course, the complete mapping would be written down directly. Looping the four adjacent squares simplifies the equation to

$$F = C$$

EXAMPLE 2.7

Use a Karnaugh map to simplify the Boolean expression

$$F = A\,B\,C + \bar{A}\,\bar{B}\,\bar{C} + A\,B\,\bar{C} + \bar{A}\,\bar{C}$$

Solution
The mapping of the expression is

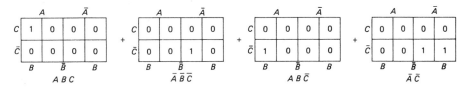

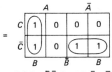

The squares can be looped together in two groups of two as shown. From this mapping,

$$F = A\,B + \bar{A}\,\bar{C}$$

The Karnaugh map can also be used to simplify an equation of the product-of-sums form as shown by the next example.

EXAMPLE 2.8

Use a Karnaugh map to simplify the equation

$$F = (A\,C + A\,\bar{C}D)(A\,D + A\,C + B\,C)$$

Solution
The mappings of $(A\,C + A\,\bar{C}D)$ and $(A\,D + A\,C + B\,C)$ are

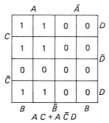 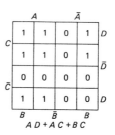

To combine the two maps to obtain the mapping of F, note that $1.1 = 1$ and $1.0 = 0$. Hence ANDing the two maps gives

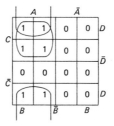

Hence

$$F = A\,C + A\,D$$

To check, using the logic rules given on page 9, first multiply out the equation to give

$$F = ACAD + ACAC + ACBC + A\bar{C}DAD$$
$$+ A\bar{C}DAC + A\bar{C}DBC$$

But $A\,A = A$ $C\,C = C$ $\bar{C}C = 0$ and so

$$F = ACD + AC + ABC + A\bar{C}D$$
$$= A\,D(C + \bar{C}) + A\,C(1 + B)$$
$$= A\,D + A\,C$$

Designing a Circuit from a Truth Table

If the truth table of a required logical operation is written down, it can be used to derive an expression for the output signal of the necessary circuit. This, after suitable simplification, can be implemented by the interconnection of a number of appropriate gates. Each 1 appearing in the output column of the truth table must be represented by a term in the

Boolean equation describing the circuit. This term must contain each variable that is at 1 and the complement of each variable that is at 0.

As an example of the technique consider the truth table given by Table 2.9.

Table 2.9

A	0	0	0	0	0	0	0	1	1	1	1	1	1	1	1
B	0	0	0	1	1	1	1	0	0	0	0	1	1	1	1
C	0	0	1	0	0	1	1	0	0	1	1	0	0	1	1
D	0	1	0	1	0	1	0	1	0	1	0	1	0	1	0
F	0	1	0	0	1	0	0	1	1	1	1	0	1	0	0

The Boolean expression for this circuit is

$$F = \bar{A}\,\bar{B}\,\bar{C}\,D + \bar{A}\,B\,\bar{C}\,\bar{D} + A\,\bar{B}\,\bar{C}\,D + A\,\bar{B}\,\bar{C}\,\bar{D} + A\,\bar{B}\,C\,D$$
$$+ A\,\bar{B}\,C\,\bar{D} + A\,B\,\bar{C}\,\bar{D}$$

The Karnaugh mapping of the function F is shown with adjacent squares looped together. From the map the simplified equation representing the logical operation given by the truth table is

$$F = A\,\bar{B} + B\,\bar{C}\,\bar{D} + \bar{B}\,\bar{C}\,D$$

Another example is the design of a **half-adder**; this is a circuit that adds two inputs A and B to produce a SUM and a CARRY but cannot take account of any carry originating from a previous stage. The truth table of a half-adder is given by Table 2.10.

Two different equations can be obtained from the truth table to express the sum and the carry of A and B. Thus

$$S = A\,\bar{B} + \bar{A}\,B \qquad C = A\,B$$

or

$$S = (A + B)\overline{AB} \qquad C = A\,B$$

(Karnaugh map)

	A		$\bar{A}$		
C	0	1	0	0	D
	0	1	0	0	$\bar{D}$
	1	1	0	1	
$\bar{C}$	0	1	1	0	D
	B	$\bar{B}$	B		

Table 2.10 Half-adder

A	0	1	0	1
B	0	0	1	1
Sum	0	1	1	0
Carry	0	0	0	1

Fig. 2.15 Two half-adder circuits

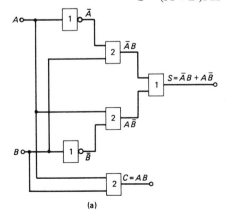

(a)

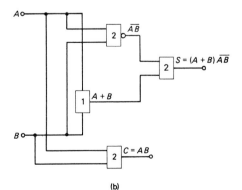

(b)

Implementing these equations using AND or OR gates gives the logic circuits shown in Figs. 2.15a and b respectively. Clearly the second circuit is more economical in its use of gates.

The first step in obtaining the NAND versions of these circuits is to replace each AND and each OR gate by its NAND equivalent. This has been done for the Fig. 2.15a circuit and the result is given in Fig. 2.16a. The circuit can then be simplified by the removal of redundant gates to give the circuit of Fig. 2.16b. Similarly, the NOR gate version of the 2.15a circuit is shown by Fig. 2.16c. It can be seen that in this case there are four redundant gates. Obtaining the NAND and the NOR equivalents of the second half-adder circuit is left as an exercise (see Exercises 2.7 and 2.17).

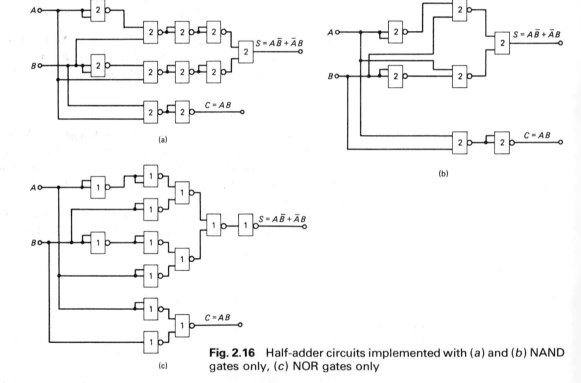

Fig. 2.16 Half-adder circuits implemented with (a) and (b) NAND gates only, (c) NOR gates only

Comparing the half-adder circuits shown in Figs. 2.16b and c, it is clear that the NOR version requires two more gates than the NAND equivalent. This is an indication that the NAND gate implementation of logical functions is more suited to those functions in sum-of-products form. Conversely, NOR gate implementation is the most appropriate for a function in product-of-sums form.

Table 2.11 Full adder

A	0	1	0	0	1	1	0	1
B	0	0	1	0	1	0	1	1
C_{in}	0	0	0	1	0	1	1	1
Sum	0	1	1	1	0	0	0	1
C_{out}	0	0	0	0	1	1	1	1

Table 2.11 gives the truth table of a **full adder**; this circuit adds together two inputs A and B and a carry from a previous stage to produce a SUM and a CARRY.

From the truth table,

$$S = A\,\bar{B}\,\bar{C}_{in} + \bar{A}\,B\,\bar{C}_{in} + \bar{A}\,\bar{B}\,C_{in} + A\,B\,C_{in}$$
$$= (A\,\bar{B} + \bar{A}\,B)\bar{C}_{in} + (\bar{A}\,\bar{B} + A\,B)C_{in}$$
$$= (\text{Exclusive-OR})\bar{C}_{in} + (\text{Exclusive-NOR})C_{in}$$

Also from the truth table.

$$C_{out} = A\,B\,\bar{C}_{in} + A\,\bar{B}\,C_{in} + \bar{A}\,B\,C_{in} + A\,B\,C_{in}$$
$$= A\,B(\bar{C}_{in} + C_{in}) + C_{in}(A\,\bar{B} + \bar{A}\,B)$$
$$= A\,B + C_{in}(\text{Exclusive-OR})$$

The sum and the carry functions can be augmented using integrated circuit exclusive-OR packages or, of course, the exclusive-OR/NOR functions can themselves be implemented

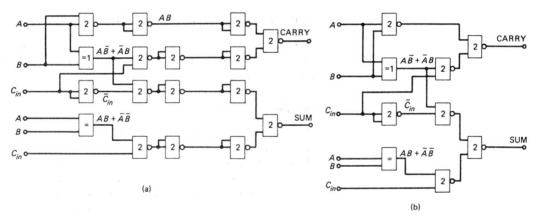

(a)   (b)

Fig. 2.17 Full adder: (a) basic circuit, (b) circuit after removal of redundant gates

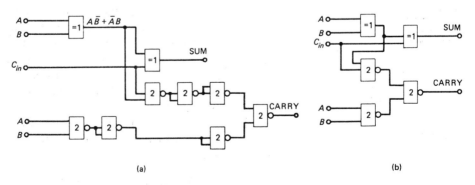

(a)   (b)

Fig. 2.18 A simpler full adder: (a) basic circuit, (b) simplified circuit

using NAND (or NOR) gates. Fig. 2.17a shows how the full adder circuit can be made using i.c. exclusive-OR and NAND gates, and Fig. 2.17b shows the circuit simplified by the removal of the redundant NAND gates. A simpler arrangement is given in Fig. 2.18a, and the simplified version by Fig. 2.18b. It is not obvious that this circuit achieves the desired result but the output of the second exclusive-OR gate is

$$F = \bar{C}_{in}(A\,\bar{B} + \bar{A}\,B) + C_{in}\overline{(A\,\bar{B} + \bar{A}\,B)}$$

$$= \bar{C}_{in}(A\,\bar{B} + \bar{A}\,B) + C_{in}(\overline{A\bar{B}} \cdot \overline{\bar{A}\,B})$$

$$= \bar{C}_{in}(A\,\bar{B} + \bar{A}\,B) + C_{in}(\bar{A} + B)(A + \bar{B})$$

$$= \bar{C}_{in}(A\,\bar{B} + B\,\bar{A}) + C_{in}(A\,B + \bar{A}\,\bar{B})$$

which is, of course, the expression for the SUM function.

Complete full adder circuits are available as integrated circuits, for example the t.t.l. 7482.

Further Applications of Logic Gates

The number of possible applications of logic gates is extremely large and in this final section of the chapter some simple examples are given.

(a) Door Alarm

A system is required which will allow a door to be opened only when the correct combination of four push-buttons is pressed. Any incorrect combination is required to bring up an alarm.

Let the four buttons be labelled as A, B, C and D and suppose that the correct combination to open the door is $A = 1$, $B = 1$, $C = 0$ and $D = 0$. Then the output F of the system which opens the door is

$$F = A\,B\,\bar{C}\,\bar{D}$$

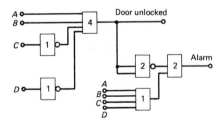

Fig. 2.19 Door alarm circuit

and the system can be implemented using a 4-input AND gate and two NOT gates as shown by Fig. 2.19. The alarm requirement is easily satisfied by connecting the output of the 4-input AND gate via a NOT gate to the alarm circuit. Further circuitry would be needed to ensure that the alarm did not operate continually, and hence the NOT gate output is fed into an AND gate along with the output of an OR gate whose inputs are the four push-buttons. The alarm will now operate only when A or B or C or D and the output of the NOT gate are at 1.

(b) High Voltage Power Supply

Access (A) to the high-voltage section of a radio transmitter should only be possible if (i) the power (P) has been switched off, (ii) the door to the section has been unlocked with a special key (D) to ensure that the power cannot be turned on again by another person, (iii) the h.t. line (H) has been earthed.

The Boolean expression describing this action is

$$A = \bar{P} D H$$

and this is easily implemented using one 3-input AND gate and one NOT gate.

(c) Motor Control

An electrical motor is to operate when (i) the power supply is connected (P), (ii) the current taken from the supply is less than some safety factor figure (I), (iii) the power supply should not be able to be switched on unless a safety guard (S) is in position, although this can be overridden by a maintenance technician by means of a special key (K).

The required logical function is

$$F = P I S \bar{K} + P I \bar{S} K$$

(d) Self-service Petrol Pump

A self-service petrol pump is to provide the required grade of petrol (P) if the pump is switched on (S), and the grade selector is positioned to one of the 4-star, 5-star and 2-star positions ($4, 5, 2$) and a button (B) is pressed to alert the cashier that the pump is in operation.

The Boolean expression representing this action is

$$P = S B (4 + 5 + 2)$$

and this can be implemented by one 3-input AND gate and one 3-input OR gate.

(e) Machine Control

An electrically controlled machine should only operate (O) if the power supply is switched on (P), a safety guard is in place (G), either manual (M) or automatic (A) operation has been selected, and, lastly, fast, medium or slow speed has been chosen (F or N or S).

The logical operation of the machine is

$$O = P G (M + A)(F + N + S)$$

Implementation requires one 2-input OR gate, one 3-input OR gate, and one 4-input AND gate.

Exercises

2.1. (*a*) Give the advantages of using only one type of gate in a logic circuit.

(*b*) With the aid of a logic diagram show how the function

$$F = \bar{A} + B\,C\,D + \overline{E\,F\,G}$$

can be carried out using only a series of 3-input NOR gates. The input and output signals at each NOR gate must be indicated and any calculations or diagrams used in obtaining the solution must be shown. (*C & G*)

2.2. (*a*) Using AND, OR and NOT gates draw a logic diagram of a circuit capable of detecting an exclusive-OR situation.

(*b*) Verify the function by drawing up the truth tables for each logic element used.

(*c*) How could the logic diagram be modified to produce an equivalence (coincidence) circuit? (*C & G*)

2.3. Write down the Boolean equation that represents the logic diagram given in Fig. 2.20. Redraw the circuit using either NAND or NOR gates only.

2.4. Fig. 2.21 shows a logic circuit. Determine the combinations of the inputs *A*, *B*, *C* and *D* that will produce a logic 1 at the output *F*. Redraw the circuit using either NAND or NOR gates only.

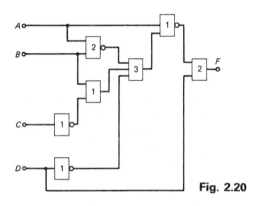

Fig. 2.20

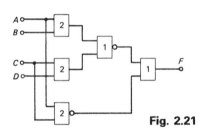

Fig. 2.21

2.5. Fig. 2.22 shows four time-related waveforms obtained from a series of logic circuits. Show three separate logic arrangements for obtaining the waveforms *E*, *F* and *G*. (*part C & G*)

2.6. Determine Boolean expressions for the outputs F_1 and F_2 of Fig. 2.23. Hence say what sort of circuit is shown.

2.7. Obtain the NAND gate version of the half-adder circuit given in Fig. 2.15*b*.

2.8. From the truth table shown, determine the Boolean equation for the output *F*. Simplify the equation using a Karnaugh map and implement the circuit using either NAND or NOR gates only.

A	0	0	1	1	0	1	1	0
B	0	0	1	0	1	1	0	1
C	1	0	1	0	0	0	1	1
F	0	0	1	1	0	0	1	0

Fig. 2.22

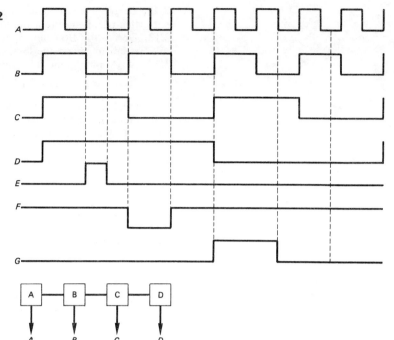

Fig. 2.23

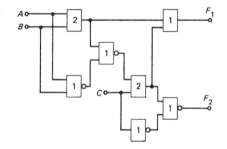

2.9. Simplify, using the rules of Boolean algebra, each of the following:
  (i) $(A+B)AB$
  (ii) $\bar{A}+\bar{C}(A+B)+C$
  (iii) $AB(\bar{A}+\bar{B}+\bar{C})$
  (iv) $AC+A\bar{C}+\bar{A}C+A\bar{A}+B+\bar{B}$

2.10. Simplify, using a Karnaugh map, each of the following:
  (i) $A+B+AC+BC$
  (ii) $AC+ABC+ABCD+ABD$
  (iii) $A\bar{C}+\bar{A}BC+A\bar{B}C\bar{D}+\bar{A}\bar{B}\bar{D}$
  (iv) $A\bar{B}C\bar{D}+\bar{A}B\bar{C}\bar{D}+AB\bar{C}\bar{D}$

2.11. Simplify, using a Karnaugh map, each of the following:
  (i) $(A+B)(C+D)$
  (ii) $(A+BC)(A+C+D)$
  (iii) $(AB+BC)(AC+ABD+BCD)$

2.12. Derive the Boolean expressions for a half-adder and use this result to obtain the logical diagram of a half-adder.

2.13. Derive the Boolean expressions for a full adder and use this result to obtain the logical diagram of a full adder.

2.14. (*a*) Minimize the following Boolean expression

$$F = A B C D + \bar{A} \bar{C} \bar{D} + A \bar{B} D + A \bar{C} D + \bar{A} C \bar{D}$$

(*b*) Draw a logic diagram using AND, OR and NOT logic elements or the minimized function.

(*c*) What single logic symbol could replace the logic diagram?

(*C & G*)

2.15. (*a*) Using Boolean algebra, minimize the following expression

$$F = A C + A \bar{C} + \bar{A} B + B \bar{B}$$

(*b*) Using the expression in (*a*) show how a map method produces the same minimized result. (*part C & G*)

2.16. (*a*) Use a mapping method to minimize the following Boolean expression

$$F = A \bar{B} C D + A \bar{B} C \bar{D} + \bar{A} B \bar{C} \bar{D} + \bar{A} B \bar{C} D + \bar{A} \bar{B} \bar{C} D$$

(*b*) Draw a logic diagram to show how NAND gates only may be used to represent the minimized expression. (*C & G*)

2.17. Obtain the NOR equivalent of the half-adder circuit shown in Fig. 2.15*b*.

Short Exercises

2.18. Show how the NAND logical function can be implemented using NOR gates.

2.19. Show how the NOR logical function can be implemented using NAND gates.

2.20. Write down the Boolean equations for the circuits shown in Figs. 2.24*a* and *b*.

2.21. Construct the truth table for (*a*) a 4-input NAND gate, (*b*) an 8-input NOR gate.

2.22. Write down the Boolean equation for a 4-input NAND gate. Show how it could be used to provide the function

$$F = \bar{A} + \bar{B} + \bar{C} + \bar{D}$$

2.23. Write down the Boolean equation for a 4-input NOR gate. Show how it could be used to provide the function

$$F = \bar{A} \bar{B} \bar{C} \bar{D}$$

2.24. Determine the necessary logical states of the inputs *A*, *B*, *C* and *D* for the output *F* of the circuit given in Fig. 2.25 to be 1.

2.25. Explain what is meant by the terms *enable* and *inhibit* when applied to logic gates.

2.26. Show how the AND and the OR logical functions can be obtained using NAND or NOR gates only.

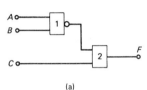

(a)

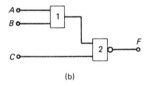

(b)

Fig. 2.24

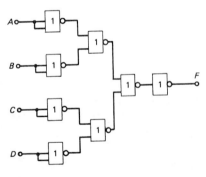

Fig. 2.25

3 Practical Electronic Gates

The various kinds of electronic gate described in the previous chapter are manufactured in a number of standard types. Some of the methods of making an electronic gate circuit are only suitable for use with discrete circuitry and so are rarely employed in modern equipment, whilst others are eminently suitable for implementation in integrated circuitry. The simplest logic family was introduced in a previous volume [EII] and is known as diode-resistor logic. Diode-resistor logic suffers from a number of drawbacks that can be overcome by the use of an active element such as a bipolar or a field effect transistor. Resistor-transistor and diode-transistor logic can also be fabricated with discrete components. When digital integrated circuits became possible, the resistor-transistor logic family was the first to be fabricated in integrated form and was followed by the diode-transistor family. These early families of integrated logic circuits have since been replaced by the **transistor-transistor** (t.t.l.), **emitter-coupled transistor** (e.c.l.), and **complementary metal oxide semiconductor** (c.m.o.s.) logic families. The most popular and most widely used circuits in modern digital equipment are in the t.t.l. and c.m.o.s. families. Diode-transistor logic is still available from some manufacturers but it is rarely, if ever, used in new equipments, while e.c.l. is only used when its special features, i.e. its very fast speed of operation, is of particular interest. The emphasis throughout this book will be on the t.t.l. and the c.m.o.s. integrated logic families.

Requirements of a Logic Gate

The various families of logic gates possess different characteristics, which mean that one of them may prove to be best suited for a particular application. For example, it may well be that the most important consideration in a particular system is the maximum possible speed of operation; for another system minimum power dissipation might be of overriding importance. The characteristics of the logic families can be classified under the following headings:

(i) Speed of operation
(ii) Fan-in
(iii) Fan-out
(iv) Noise margin or noise immunity
(v) Power dissipation.

Speed of Operation

The **speed of operation** of a logic gate, also commonly known as its **delay time** or **propagation time**, is the time that elapses between the application of a signal to an input terminal and the resulting change in logical state at the output terminals. The delay arises because the output voltage of a switching transistor is unable to change instantaneously from one logic value to another when its input voltage is changed. Suppose

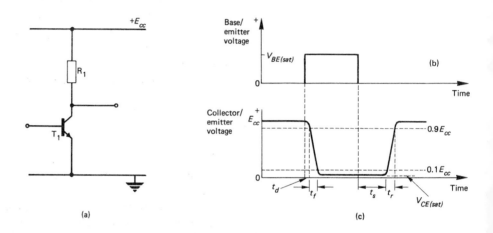

(a)   (c)

Fig. 3.1 Switching a bipolar transistor circuit: (a) basic circuit, (b) input voltage, (c) output voltage

the bipolar transistor circuit given in Fig. 3.1a has the rectangular voltage of Fig. 3.1b applied to its base terminal. When the base voltage is zero the transistor is non-conducting or OFF and its collector/emitter voltage V_{CE} is equal to the collector supply voltage E_{cc}. When the base voltage is increased in the positive direction, the transistor starts to conduct current and its collector/emitter voltage falls because of the voltage dropped across the collector load resistor R_L, i.e.

$$V_{CE} = E_{cc} - I_c R_L$$

The collector/emitter voltage does not start to fall at the same instant as the base/emitter voltage goes positive (see Fig. 3.1) but after a time t_d has elapsed.

This time delay is present because of the need for the base current to charge the base/emitter junction capacitance, and because of the time taken for charge carriers to cross the base region. Once the output voltage commences to fall it takes a finite time to do so because the collector/base junction capacitance must also be charged. The **fall-time** t_f is the time taken for the output voltage to fall from 90% to 10% of its OFF value, E_{cc}. When the transistor is fully conducting, or *saturated*, it is said to be ON and then its collector/emitter voltage has fallen to its minimum value, known as the *saturation voltage* $V_{CE(sat)}$. Once the transistor has saturated, any further increase in the base current cannot produce a further increase in the collector current, since this is limited to $(E_{cc} - V_{CE(sat)})/R_L$, and the excess charge is stored in the base region of the transistor.

When the base voltage of the transistor is reduced to zero the transistor does not immediately switch off; instead there is a time delay t_s before the collector/emitter voltage starts to increase to its OFF value of $+E_{cc}$ volts. The delay, known as the **storage delay**, is caused by the need for the excess charge stored in the base region to be removed before the collector current can change value. Once the collector/emitter voltage does start to increase, it takes a time, known as the **rise time** t_r, to rise from 10% to 90% of its final (OFF) value. This delay arises from the time taken for the collector current to change between its ON and OFF states.

Typically, the total ON and OFF times are about 6 ns and 10 ns respectively and to increase the switching speed the transistor must be prevented from driving into saturation. This can easily be achieved by the connection of a diode between the base and the collector terminals of the transistor as shown by Fig. 3.2. When the transistor is turned ON by a positive voltage applied to its base, its collector/emitter voltage falls and immediately the collector is less positive than the base, diode D_1 conducts and diverts the excess base current away from the base. As a result the transistor does not receive sufficient base current to be driven into saturation and charge storage is no longer a problem, It is best if the diode is of the Schottky type since these devices have zero charge storage and so are very fast switching. A **Schottky transistor** is a bipolar transistor which has a Schottky diode internally connected between its base and collector terminals (Fig. 3.3a). The symbol for a Schottky transistor is given in Fig. 3.3b.

The **propagation delay** of an electronic gate not only takes account of the switching times of a transistor but also the facts that (i) the input voltage pulse waveform will not be perfectly rectangular but will itself exhibit rise and fall times and (ii) the input voltage need only reach some threshold value before the circuit starts to switch. The propagation delay is the time required for the output voltage of a gate to change after the application of an input voltage. Values of propagation delay vary considerably with the type of circuit but for integrated circuits they are normally 5–40 ns per gate.

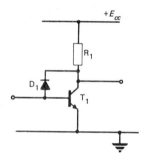

Fig. 3.2 Use of a diode to increase switching speed

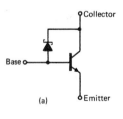

(a)

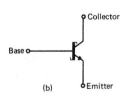

(b)

Fig. 3.3 (a) Schottky transistor, (b) symbol

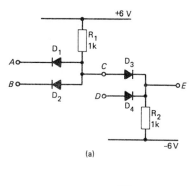

(a)

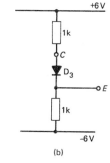

(b)

Fig. 3.4

Fan-in and Fan-out

The **fan-in** of a gate is the number of inputs, coming from similar circuits, that can be connected to the gate without adversely affecting its performance.

The **fan-out** of a gate is the maximum number of similar circuits that can be connected to its output terminals without the output voltage falling outside the limits at which the logic levels 1 and 0 are specified.

As an example consider the diode logic arrangement shown in Fig. 3.4a; an AND gate having a fan-in of 2 has its output terminal C connected to one of the inputs of an OR gate. When both inputs A and B are at zero volts, logical 0, the voltage at point C differs from 0 V only by the small voltage dropped across the diodes. Neglecting this voltage drop, the input C to the OR gate is at 0 V. If the input D is at +6 V, logical 1, the output E will be at +6 V and the circuit works correctly. When the condition $A = B = 1$ arises, the voltage at C should be equal to +6 V but this voltage will be reduced by the input impedance of the OR gate. The circuit is redrawn as shown in Fig. 3.4b. Neglecting the small voltage dropped across the diode the voltage at C is equal to

$$1 \times 12/(1+1)-6=0 \text{ V}$$

Thus the output of the AND gate is at logic 0 instead of at logic 1. For correct operation of the circuit it is necessary to increase the value of the resistor R_2. Suppose R_2 is made 10 kΩ. Then the voltage at C, which is also the voltage at E, is equal to

$$10 \times 12/(1+1)-6=4.9 \text{ V}$$

which will be taken as logic 1. The requirement for correct operation of the circuit is therefore that the *value of the gate resistor R_1 must be very much less than the load resistance applied to the gate.*

The resistance of the loads applied to a gate's output terminals is also a factor in determining the fan-out of a diode gate. Suppose that a second OR gate is connected to the output C of the AND gate of Fig. 3.4a. The resistance connected across the output of the AND gate will then be 10 kΩ in parallel with 10 kΩ or 5 kΩ. This will make the voltage at C fall to

$$5 \times 12/(5+1)-6=4 \text{ V}$$

This voltage may still be large enough to be representative of logic 1 but clearly the addition of further gates will soon reduce the load resistance to the point at which a logic 1 output cannot be obtained. Thus, *the higher the resistance of the gate load, the greater the fan-out of a gate.*

If a diode gate is overloaded, i.e. its fan-out is exceeded, it will not be possible to obtain the logic 1 state at its output terminals.

Power Dissipation

Power is dissipated within a transistor as it switches from one state to another and also within all current-carrying resistors. The greater the power dissipation of a gate, the more heat must be removed from the circuit; also, particularly if many gates are used within an equipment, larger and hence more costly power supplies will be necessary.

The **d.c. power dissipation** of a gate is the product of the d.c. power supply voltage and the mean current taken from that supply.

Noise Immunity or Margin

Noise is the general term for any unwanted voltages that appear at the input to a gate. The possible sources of noise are many and are discussed elsewhere [EIII]. If the noise voltage has a sufficiently large amplitude, it may cause the gate to change its output state even though the input signal voltage has remained constant. Such false operation of a gate will lead to errors in the circuit performance. The **noise immunity** or **noise margin** of a gate is the maximum noise voltage that can appear at its input terminals without producing a change in the output state. Usually, manufacturers of integrated circuit gates quote d.c. values of noise margin, giving both typical and worst case values.

The **threshold value** of a gate is the input voltage at which a change of the output state of the circuit is just triggered. A reasonable approximation to this value is the voltage midway between the two logic levels. For the t.t.l. logic family the threshold voltage is 1.4 V but the maximum input voltage that will definitely be read as logic 0 is 800 mV, whilst the minimum input voltage giving a definite logic 1 is 2.0 V.

The **logic levels** of a gate are the voltages that must exist for the circuit to operate correctly. Thus, for a t.t.l. gate, logic 1 has a typical voltage of 3.3 V and a minimum voltage of 2.4 V; on the other hand logic 0 is typically 0.2 V but maximally 0.4 V.

The noise margin of a gate is equal to the difference between the logic level at the output of the gate and the threshold value of the gate(s) to which its output is connected. This is shown by Fig. 3.5 which refers to t.t.l. NAND gates. In Fig. 3.5 it is supposed that binary 0 is applied to the inputs of the first NAND gate so that its output is at binary 1. This

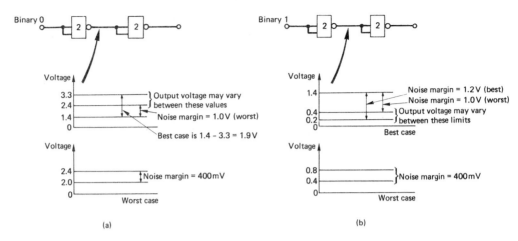

Fig. 3.5 Noise margin of t.t.l. gates

means that the output voltage lies within the limits of 2.4 V and 3.3 V. The threshold voltage is taken as 1.4 V and so the noise margin varies between 1.0 V at the worst and 1.9 V at best. The maximum value of the threshold voltage is 2.0 V, and should this exist the worst-case noise margin will be only 400 mV.

When the output of the first gate is at binary 0 its voltage will be within the range 0.2 V–0.4 V. The threshold voltage is 1.4 V and so the noise margin varies from 1.0 V at worst to 1.2 V. The minimum threshold voltage is only 0.8 V and hence the worst-case noise margin is 400 mV.

Diode-Transistor Logic

Diode-transistor logic (d.t.l.) is easily made with discrete components and is still available in integrated circuit form, although for new work it has been superseded by the t.t.l. and c.m.o.s. logic families.

The basic gate in the d.t.l. family performs the NAND function and is shown in Fig. 3.6. When both inputs A and B have positive voltages applied to them, neither diode D_1 nor diode D_2 conducts and transistor T_1 is turned ON by the base current provided by resistor R_1. The output voltage of the circuit is now very nearly zero, being equal to the saturation voltage of T_1. If either or both inputs are at 0 V, the associated diode will conduct and the base terminal of T_1 will fall to the ON potential of the diode(s). This voltage will be insufficiently positive to allow T_1 to conduct and the device will turn OFF. The output voltage of the circuit will then be $+E_{cc}$ volts. Thus, if positive logic [EII] is used, the circuit performs the NAND

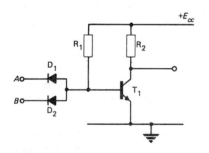

Fig. 3.6 D.T.L. NAND gate

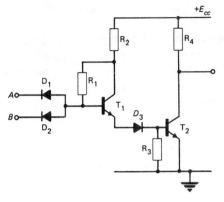

Fig. 3.7 Improved d.t.l. NAND gate

logical function. A better performance can be obtained if the gate circuitry is modified as shown by Fig. 3.7.

The input diodes D_1 and D_2 will be reverse biased and hence non-conducting when both inputs A and B have a positive voltage applied to them. T_1 is then conducting and its emitter current turns T_2 full ON. The output voltage of the circuit has then fallen to the saturation voltage $V_{CE(sat)}$ of T_2 and this is the logical 0 state. If either or both of the inputs A and B are at 0 V (logic 0), the associated diode conducts and the base voltage of T_1 becomes less positive than the value needed to keep T_1 conducting. T_2 turns OFF and its collector/emitter voltage rises to the supply voltage $+E_{cc}$ volts, or logic 1. Thus, this circuit also performs the NAND function.

The diode D_3 connected between the emitter of T_1 and the base of T_2 is provided to increase the noise margin of the circuit. Noise voltages will only appear at the base of T_2 when T_1 is OFF if their amplitude is greater than the turn-on voltage of the diode.

The fan-out of a d.t.l. gate is 8; if this figure is exceeded, the effective value of the collector load of T_2 will be reduced to such a low value that T_2 may not saturate with its maximum base drive. Suppose, for example, that the saturation voltage of T_2 is 0.2 V, the collector supply voltage is 6 V and the gate load resistance is $1000\,\Omega$. The collector current needed for saturation to occur is $(6-0.2)/1000 = 5.8\,\text{mA}$, and if the transistor has a current gain of $h_{FE} = 60$ the current which turns the transistor full ON is $5.8 \times 10^{-3}/60 = 96.7\,\mu\text{A}$. If the gate has a fan-out of 1 and the next gate has an input resistance of $1000\,\Omega$, the effective gate resistance is $500\,\Omega$ and the collector current for saturation must be 11.6 mA. For a fan-out of 4, the effective gate resistance falls to $250\,\Omega$ and $I_{c(sat)}$ is 23.2 mA. The maximum collector current is equal to $h_{FE} \times I_{b(max)}$ and this clearly sets a limit to the maximum fan-out. Generally, a discrete component d.t.l. gate has a fan-in of 14 and a fan-out of 8, but an i.c. version has, typically, fan-in and fan-out figures of 8 and 5 respectively.

Some examples of i.c. d.t.l. gates are dual 4-input NAND, quadruple 2-input NAND, quadruple 2-input OR, and quadruple 2-input AND. Typical figures for an i.c. d.t.l. gate are (i) transmission delay 30 ns, (ii) noise immunity 1 V, (iii) power dissipation 8.5 mW per gate.

Resistor-Transistor Logic

The resistor-transistor (r.t.l.) logic family was commonly used before integrated circuits became readily available but nowadays its use is fairly rare. A few versions were made in i.c.

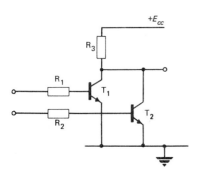

Fig. 3.8 R.T.L. NOR gate

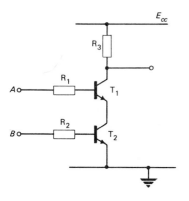

Fig. 3.9 R.T.L. NAND gate

form but they were rapidly replaced by d.t.l. and other logic families.

Fig. 3.8 shows the circuit of an r.t.l. NOR gate. When both inputs are at zero volts, or logic 0, both transistors T_1 and T_2 are turned OFF and so the output terminal is at $+E_{cc}$ volts or logic 1. If either or both input terminals have $+E_{cc}$ volts applied, the associated transistor will be turned ON and the output voltage will fall to very nearly zero volts, i.e. the output will be at logic 0. The output is at logic 1 only when both inputs are at logic 0, i.e. the NOR function is performed.

The NAND gate can also be implemented in r.t.l. and Fig. 3.9 gives a possible circuit. Only if both transistors are ON will the output voltage be approximately zero volts or logic 0. Since each transistor will only conduct when a positive voltage is applied to its associated input terminal, the NAND function is performed.

The r.t.l. logic family is fairly slow to operate but it has a good noise immunity; typically the fan-in is 4 and the fan-out is 6.

Transistor-Transistor Logic

The most popular and most widely used logic family is the transistor-transistor or t.t.l. family, which is manufactured in integrated circuit form by most semiconductor manufacturers. The popularity of this logic family arises because it offers fairly high speed (particularly the Schottky versions), good fan-in and fan-out, and is easily interconnected or *interfaced* with other digital circuitry—all at a relatively low cost. The standard t.t.l. logic, known as the 54/74 series, has a poor noise immunity and a rather high power consumption. The 74 series are designed for commercial applications and operate at ambient temperatures of up to 70°C. The 54 series is primarily intended for military circuitry and has a maximum ambient temperature of 125°C. Low-power versions of the t.t.l. gates are also available. Gates made by different manufacturers may well have different internal circuits but any two devices with the same number will perform exactly the same logical function and be pin-for-pin compatible.

NAND Gate

The basic circuit of a t.t.l. NAND gate is given in Fig. 3.10. The input transistor has a number of emitters equal to the desired fan-in of the circuit; in the figure, a fan-in of 2 has been assumed. In the 54/74 series, fan-ins of 2, 3, 4 and 8 are available. When both input terminals are at +5 V the emitter/base junction of T_1 is reverse biased but its collector/base junction is forward biased. Current then flows from the collector power supply, through R_1, into the base of T_2. Transistor T_2 turns full ON and the output voltage of the circuit falls to the saturation voltage of the transistor. The output of the circuit is then at logical 0. When either or both of the input terminals is at approximately zero volts, logic 0, the associated emitter/base junction will be forward biased (its base is more positive than its emitter). The value of resistor R_1 is selected to ensure that T_1 is then full ON and so the base voltage of T_1 is only V_{BE1} volts (approximately 0.7 V) above earth potential. This potential is insufficient to keep T_2 ON and so T_2 turns OFF. The output voltage of the circuit then rises to +5 V, i.e. becomes logical 1. Thus, transistor T_1 performs the AND function and T_2 acts as an invertor to give an overall circuit rendering of the NAND function.

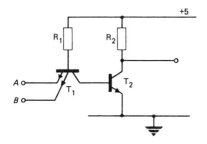

Fig. 3.10 Basic t.t.l. NAND gate

Fig. 3.11 Standard t.t.l. NAND gate

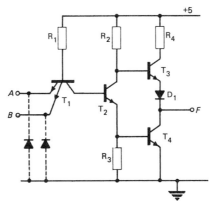

The standard t.t.l. gate adds an output stage to the basic circuit of Fig. 3.10 in order to increase both the operating speed and the fan-out, the complete circuit being given in Fig. 3.11. The output stage, consisting of transistors T_3 and T_4 and diode D_1, is often known as a *totem-pole* stage. When T_2 has turned ON the base/emitter potential of T_3 is approximately zero and so T_3 does not conduct. At the same time T_4 is turned ON by the voltage developed across resistor R_3. Thus, when T_2 is ON transistor T_3 is OFF and T_4 is ON; this means that the potential at the output terminal of the circuit is low and so the output state is logical 0. The fan-out can be up to about 10 without the saturation voltage of T_4 rising above the 0 level. Similarly, when T_2 is OFF its collector voltage is +5 V and its emitter voltage is 0 V. Now T_4 is turned OFF and T_3 conducts to an extent that is determined by the external load connected to the output terminals of the circuit. The output voltage is then equal to 5 V minus the voltage dropped in T_3 and D_1, i.e. logic 1.

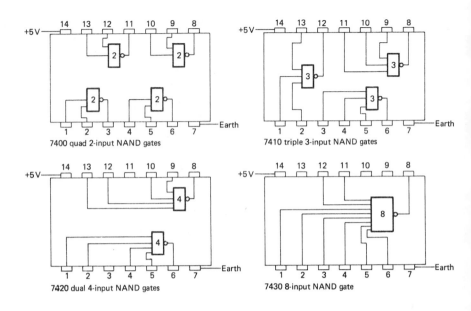

Fig. 3.12 Pin connections of four t.t.l. i.c. gates

To protect the circuit from damage by any negative voltages arriving at the input terminals, and also to improve the noise immunity, a diode can be connected between each input and earth, shown dotted in Fig. 3.11.

Many i.c. packages contain more than one gate and some examples have their pin connections shown by Fig. 3.12. The various t.t.l. gates available are listed in Appendix 1.

NOR Gate

The circuit diagram of a t.t.l. NOR gate is given by Fig. 3.13. If either T_2 or T_3 is turned ON, the base potential of T_5 will fall to very nearly 0 V and T_5 will turn OFF. Since either T_2 or T_3 is conducting heavily, the positive voltage developed across R_3 will be large enough to turn T_6 ON; then the output of the circuit will be at logical 0. The output of the circuit will only be at logic 1 when both T_2 and T_3 are OFF; then the the base potential of T_5 is +5 V and T_5 turns full ON, while T_6 is turned OFF because the voltage across R_3 is zero. Transistors T_2 and T_3 will be turned ON by a positive voltage at their base terminal and this will happen only if T_1 and/or T_4 is conducting, i.e. when input A and/or B is at +5 V or logical 1. The circuit therefore performs the NOR function since the output is at logical 1 only when both inputs A and B are at logical 0.

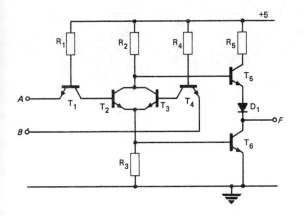

Fig. 3.13 T.T.L. NOR gate

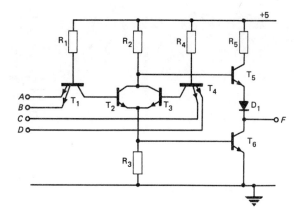

Fig. 3.14 T.T.L. AND-OR-INVERT gate

AND-OR-INVERT Gate

The circuit of a t.t.l. AND-OR-INVERT (AOI) gate is shown in Fig. 3.14. It can be seen to be very similar to the NOR gate, differing only in that the input transistors T_1 and T_4 have multiple emitters (two are shown). The operation of the circuit to produce the logical function

$$F = \overline{A\,B + C\,D}$$

is left as an exercise (Exercise 3.25). AOI gates are often described as *wide*. "Wide" means the number of inputs to the NOR gate. A dual 2-input 2-wide AOI gate (7451) performs the functions

$$F = \overline{(A\,B\,C) + (D\,E\,F)}$$

and

$$F = \overline{(A\,B) + (C\,D)}$$

while the 4-wide 2-input AOI gate (7454) performs

$$F = \overline{(A\,B) + (C\,D\,E) + (F\,G\,H) + (I\,J)}$$

A 2-wide 4-input AOI gate (7455) gives

$$F = \overline{(A\,B\,C\,D) + (E\,F\,G\,H)}$$

AND Gate

The logical function AND can be performed in the t.t.l. family by putting an inverting stage in between the NAND gate proper and the totem-pole output stage (see Fig. 3.15). The necessary signal inversion is provided by transistor T_4. The extra stage means that the operating speed is lower and the power dissipation is higher than a NAND or a NOR gate.

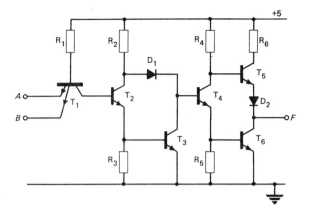

Fig. 3.15 T.T.L. AND gate

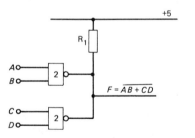

Fig. 3.16 Direct connection of open-collector gates

Open-collector Gates

Very often it is desirable to be able to connect together the outputs of several gates to increase the fan-out, or to perform a particular logical function, or perhaps to connect several gates to a common power line.

If two gates with totem pole output stages have their output terminals connected together, one of the gates is very likely to pass an excessive current which may well damage it. If, for example, the output of one gate is at logic 0, while the output of the other gate is at logic 1, the first gate would have a low resistance to earth and a current of high enough magnitude to cause damage could flow.

An **open-collector gate** is one which has been designed to permit it to be directly connected to another gate in the manner shown by Fig. 3.16. Two open-collector NAND gates have their output terminals paralleled and then connected to an external *pull-up resistor* R_1 to produce what is generally called the **wired-OR gate**. The logical function performed by a wired-OR gate is

$$F = \overline{A\,B} + \overline{C\,D}$$

If the same function is produced using totem-pole output stages, an extra gate is necessary, thus increasing the propagation delay (also the power dissipation is larger). Disadvantages of open-collector gates are: external resistor needed; slower operation; and poorer noise margin.

The connection is known as the wired-OR because, if the output of either gate goes to zero volts, the output of the paralleled gates must also become 0 V. Only if both outputs are at logical 1 can the combined output be 1. The method of connection can be extended to more than two open-collector gates. The wired-OR connection is also possible using gates in the d.t.l. family.

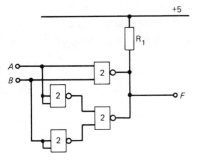

Fig. 3.17 Exclusive-OR gate using open-collector gates

The wired-OR principle can be applied to form an exclusive-OR gate, (see Fig. 3.17). For this circuit,

$$F = \overline{A\,B + \bar{A}\,\bar{B}}$$
$$= (\overline{A\,B})(\overline{\bar{A}\,\bar{B}})$$
$$= (\bar{A} + \bar{B})(A + B)$$
$$= \bar{A}\,A + \bar{A}\,B + A\,\bar{B} + B\,\bar{B}$$
$$= \bar{A}\,B + A\,\bar{B} \quad \text{the exclusive-OR function}$$

An open-collector gate differs from the totem-pole output equivalent in that the components

R_4, T_3, D_1 in Fig. 3.11
R_5, T_5, D_1 in Fig. 3.13
R_6, T_5, D_2 in Fig. 3.15

are omitted. The *pull-up resistor* (R_1 in Fig. 3.16) is necessary to ensure that the output terminal goes high when the output transistor T_4 (or T_6) turns off.

Low-power T.T.L.

The series 54L/74L uses the same circuitry as the standard series but all the resistance values are increased in order to minimize the internal power dissipation.

Schottky T.T.L.

For many applications the (typical) 9 ns propagation delay time of the standard t.t.l. gate is too high and so a higher-speed version has been introduced. This alternative, known as Schottky t.t.l., series 54S/74S, uses Schottky transistors as shown in Fig. 3.18. Excess base current is diverted from the base of the transistor by the Schottky diode which clamps the

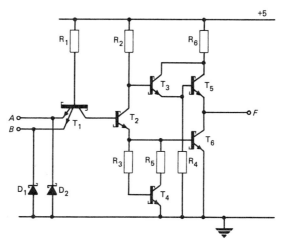

Fig. 3.18 Schottky t.t.l. NAND gate

base/collector potential to 0.4 V. Since this voltage is not large enough to drive the transistor into saturation, the switching speed is considerably increased without an accompanying increase in the power dissipation.

Differences between the NAND circuit of Fig. 3.18 and the standard gate of Fig. 3.11, other than the use of Schottky transistors, are that resistor R_3 is replaced by transistor T_4 and resistors R_3 and R_5, and D_1 is replaced by the Darlington-connected transistors T_5 and T_3 to increase the speed of operation. T_4 is often called an *active turn-off* device.

A low-power Schottky t.t.l. series is also available and is now the most commonly employed type of gate. The 54LS/74LS series is faster than the standard series and it also dissipates less power but it is slower than the 54S/74S series. For a NAND gate, for example, the different performance is achieved by using a diode input circuit as shown in Fig. 3.19.

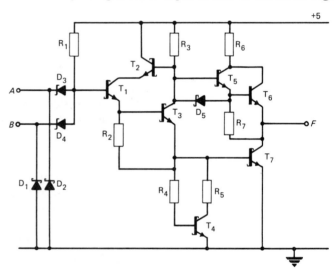

Fig. 3.19 Low-power Schottky t.t.l. NAND gate

When both inputs are at +5 V, logic 1, T_1 and T_2 conduct and cause T_3/T_4 to conduct. The collector potential of T_3 is then low and this causes T_5/T_6 to turn off. The voltage across R_4 makes T_7 conduct to take the output to logic 0.

A comparison between the various types of t.t.l. gate is given by Table 3.1, which shows typical values for the parameters quoted.

Unused Inputs

Any unused input terminals should not be left "floating," i.e. unconnected, since both the switching time and the noise immunity of the gate will be improved if unused inputs are taken, via a resistance of about 1 kΩ, to the +5 V supply line. This resistor is not necessary with low-power Schottky gates.

Table 3.1 T.T.L. gates

Type of gate	Standard	Low-power	Schottky	Low-power Schottky
Propagation delay (ns)	10	30	3	7
Noise immunity (V)	1 0	0.4 0.4	0.4 0.4	0.7 0.3	0.7 0.3
Power dissipation (mW)	10	1	20	2
Fan-out	10	10	10	10

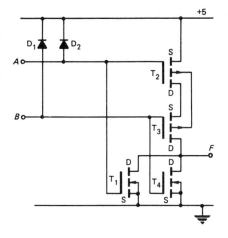

Fig. 3.20 C.M.O.S. NOR gate

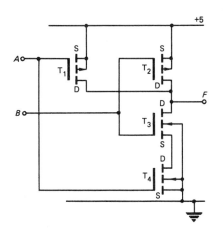

Fig. 3.21 C.M.O.S. NAND gate

Alternatively, for standard t.t.l., unused inputs can be connected together and then to a used input. For the LS series this practice reduces the noise immunity.

C.M.O.S. Logic

The complementary metal-oxide semiconductor, or c.m.o.s., logic family offers the desirable features of very low power dissipation and good noise immunity. The family finds particular application when low-power consumption is of prime importance. The main disadvantage associated with this type of gate is its limited switching speed arising from the very high input impedance of a m.o.s.f.e.t. device.

Fig. 3.20 shows the circuit diagram of a c.m.o.s. NOR gate. If either input terminal A or input terminal B or both is at $+5$ V (logic 1) the associated p-channel m.o.s.f.e.t. (T_2 and/or T_3) is turned OFF, and the associated n-channel m.o.s.f.e.t. is turned ON. The drain/source voltages of T_1 and T_4 are then approximately zero and so the output state of the circuit is logical zero. If both input terminals A and B are at logic 0, m.o.s.f.e.t.s T_1 and T_4 turn OFF. The output terminal of the circuit is now at very nearly $+5$ V and so the output state is logic 1. The output of the gate is logic 1 only when both inputs are at logic 0 and so the circuit performs the NOR function.

The circuit of the c.m.o.s. NAND gate is given in Fig. 3.21. The operation of this circuit is as follows. If either A or B or both is at zero volts, or logical zero, then m.o.s.f.e.t. T_1 and/or T_2 is turned ON but m.o.s.f.e.t. T_3 and/or T_4 is OFF. The output terminal is then at very nearly $+5$ V (logic 1). Conversely, if both A and B are at $+5$ V, T_1 and T_2 are OFF while T_3 and T_4 are ON. The output of the circuit is then at approximately zero volts or logic 0.

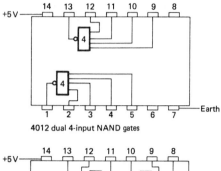

4012 dual 4-input NAND gates

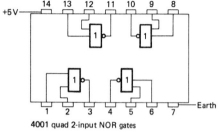

4001 quad 2-input NOR gates

Fig. 3.22 Pin connections of two c.m.o.s. packages

Many c.m.o.s. gates have protective diodes connected between their input terminals and earth to reduce the possibility of damage to the device caused by handling or soldering. The gate terminal of a m.o.s.f.e.t. is insulated from the channel by a very thin layer of insulation which effectively forms the dielectric of a capacitance. Any electric charge which accumulates on the gate terminal may easily produce a voltage that is large enough to cause the dielectric to break down. Once this happens the device has been destroyed. The charge necessary to damage a m.o.s.f.e.t. need not be very large since the capacitance between the gate and the channel is very small and $V = Q/C$. This means that a dangerously high voltage can easily be produced by merely touching the gate lead with a finger or a tool. Great care must therefore be taken when a c.m.o.s. circuit is fitted into, or is removed from, a circuit. The leads of a c.m.o.s. device in store are usually short-circuited together by springy-wire clips or by conductive jelly or grease, and the short circuit should be maintained while the device is being fitted into circuit, particularly during the soldering process. The solderer should stand on a non-conducting surface such as a rubber mat and should use a non-earthed soldering iron. The i.c. leads should be protected from heat by the use of a heat shunt (pliers) and the device should be allowed to cool after each connection is soldered before tackling another one. When an i.c. is removed from a circuit a de-soldering tool should be used to remove all the solder from the connections and then the i.c. can be lifted off from its tags. The problem is, of course, simplified if i.c. holders are used.

Table 3.2 Functionally equivalent T.T.L./C.M.O.S. gates

Function	T.T.L.	C.M.O.S.	Function	T.T.L.	C.M.O.S.
Quad 2-input NAND	7400	4011	Quad 2-input NOR	7402	4001
Quad 2-input AND	7408	4081	Triple 3-input NAND	7410	4023
Triple 3-input NOR	7411	4073	Double 4-input NAND	7420	4012

Fig. 3.22 shows the pin connections of two of the devices in the c.m.o.s. family; the pin connections shown are standard to all manufacturers. Other gates in this series are listed in Appendix 2. Unused inputs should be taken to the +5 V supply or earth depending on the logic function, or connected to another used input. Otherwise potentials may be developed which will lead to false logical operation of the gate. Gates in

the c.m.o.s. family have pin-for-pin equivalents in the t.t.l. family and are also easily interfaced with them. Table 3.2 lists some t.t.l. to c.m.o.s. functionally equivalent types of gate.

Interfacing T.T.L. and C.M.O.S.

Very often the need may arise for gates in the t.t.l. and c.m.o.s. families to be interconnected or **interfaced**. The output of a t.t.l. gate cannot be directly connected to the input of a c.m.o.s. gate without adversely affecting the noise immunity of the circuit. To overcome this problem, a **pull-up resistor** can

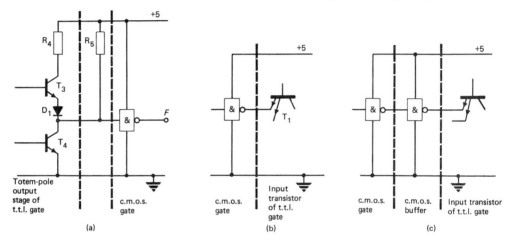

Fig. 3.23 Interfacing t.t.l. and c.m.o.s. gates

be connected between the junction of the two gates and the +5 V supply line (Fig. 3.23a). It is possible to directly connect the output of a c.m.o.s. gate to the input of a t.t.l. gate as shown by Fig. 3.23b, but generally a c.m.o.s. buffer stage, such as the 4010 is used (Fig. 3.23c).

Emitter-coupled Logic

The main feature of emitter-coupled logic, or e.c.l., is the very fast speed of operation that is provided. This logic family is used when the maximum possible speed is the prime consideration. Very fast operation is obtained by designing the circuitry to ensure that the transistors do not drive into saturation when conducting. The basic e.c.l. gate is a combined OR/NOR circuit and this, in common with the rest of the family, is operated from a −5.2 V supply. This means that logic 1 is represented by −0.9 V and logic 0 by −1.75 V.

The circuit diagram of an OR/NOR e.c.l. gate is given by Fig. 3.24. A reference voltage of −1.29 V is developed by the

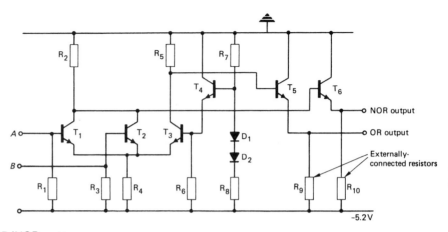

Fig. 3.24 E.C.L. OR/NOR gate

R_7, D_1, D_2, R_8 circuit; and applied to the base terminal of transistor T_4. If both inputs A and B are at logical zero (-1.75 V) then the base potential of T_3 is more positive (less negative) than the base potentials of either T_1 or T_2, and so T_3 conducts whilst T_1 and T_2 do not. The collector current of T_3 develops a voltage of about -1.0 V across R_5 and this causes T_5 to conduct. T_5 is connected as an emitter follower [EIII] and so its output voltage is $-1.0\,V - V_{BE5} \simeq -1.7\,V$ which represents logical 0. The collector potentials of T_1 and T_2 are approximately zero and hence the output voltage of T_6 is $0\,V - V_{BE6} \simeq -0.7\,V$ which represents logical 1.

If either input A or B is at logical 1, i.e. -0.9 V, the associated transistor is turned ON while T_3 is turned OFF. Now the collector potentials of T_1/T_2 and T_3 are reversed compared to the previous case and so now, T_5 output is at logical 1 and T_6 output is at logical 0.

Since each input is terminated by a resistor, an unused input need not be connected to a fixed voltage level.

When an e.c.l. circuit is to interface with c.m.o.s. or t.t.l. circuitry, the connection must be made via a *level shifting circuit* because of the -5.2 V supply voltage of e.c.l. Level shifter circuits are available in the c.m.o.s. family, e.g. 4049/50.

Comparison between Integrated Logic Families

The main characteristics of the various logic families which are available in integrated circuit form are listed in Table 3.3. Typical figures are quoted.

Systems such as mainframe computers where the highest possible speed of operation is of the utmost importance use e.c.l. but most other systems use one form or other of t.t.l. or c.m.o.s.

Table 3.3 Integrated circuit logic families

Family	Propagation delay (ns)	Power dissipation (mW)	Noise immunity (V)	Fan-out	Supply voltage (V)	Fan-in
d.t.l.	30	8	1	5	5	8
t.t.l. standard	9	40	0.4	10	5	8
t.t.l. Schottky	3	40	0.3	10	5	8
t.t.l. low-power Schottky	8	8	0.3	10	5	8
c.m.o.s.	30	.001	1.5	50	5	8
e.c.l.	1.1	30	0.4	50	−5.2	5

Recent developments have been the introduction of improved t.t.l. devices such as Advanced Schottky and Advanced Low-Power Schottky. These devices provide a much faster speed of operation together with a lower power dissipation. The speed of Advanced Schottky is such that it can compete successfully with e.c.l. for very-high speed applications.

Exercises

3.1. (*a*) With the aid of a suitable diagram explain the meaning of the following terms used in connection with transistor logic elements: (i) saturation, (ii) hole storage time.
(*b*) Draw a circuit diagram of a typical 3-input t.t.l. NAND gate, explaining clearly the method used to overcome the effects of saturation. (*C & G*)

3.2. (*a*) Draw a circuit of a 2-input diode-transistor positive logic NAND gate with an emitter follower output stage, and explain both its logical and electrical operation.
(*b*) Sketch, on the same voltage/time axes, a typical input and output waveform for the emitter follower used above.
(*c*) State the reason for using the emitter follower. (*C & G*)

3.3. Fig. 3.25 shows the circuit of a c.m.o.s. NOT gate. Explain its operation.

3.4. An exclusive-OR gate is made using standard t.t.l. NAND gates; both totem-pole output stage and open-collector versions are available. Draw an example of each.

3.5. Draw the circuit diagram of a Schottky t.t.l. 3-input NAND gate. With the aid of a truth table explain its operation.

3.6. Draw the circuit of a c.m.o.s. 2-input NAND gate and explain fully the action of the circuit.

3.7. (*a*) Describe, with aid of a circuit diagram and a truth table, the operation of a 3-input NAND gate using t.t.l. logic.
(*b*) Compare the speed, power dissipation and noise immunity of the t.t.l. family with e.c.l. and c.m.o.s.

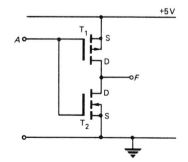

Fig. 3.25

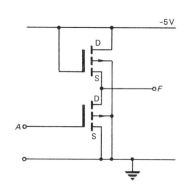

Fig. 3.26

3.8. (*a*) With the aid of a circuit diagram and a truth table describe the operation of 3-input d.t.l. NOR gate.
(*b*) Explain how the value of the collector resistor determines the fan-out capability of the gate described.
(*c*) How does the noise immunity of the gate described compare with that of a NOR gate using negative logic?   (*C & G*)

3.9. The logic function $F = A B + C D$ is to be implemented using (i) t.t.l. NAND and/or NOR gates, (ii) t.t.l. AOI gates. Draw a circuit for each case (*a*) showing each gate used as a separate block, (*b*) showing each i.c. used as a separate block. Say whether or not you are using open-collector gates.

3.10. Explain, with the aid of circuit diagrams, the difference between totem-pole and open-collector output stages for t.t.l. gates.

3.11. Fig. 3.26 shows the circuit of a c.m.o.s. gate. Describe the operation of the circuit and <u>say what</u> kind of gate it is.

3.12. Implement the function $F = \overline{(A B + C D)}A C$: (*a*) without simplification using (i) AOI and AND gates, (ii) either NAND or NOR gates; (*b*) simplify the equation and use NAND gates only.

3.13. Explain, with the aid of a diagram, the operation of a t.t.l. Schottky NAND circuit. Show how one or more NAND gates can be used to produce (i) the OR function, (ii) the NOT function.

Short Exercises

3.14. What is the advantage of using the wired-OR connection compared with using cascaded totem-pole output stage gates to obtain a given logical function?

3.15. What is a Schottky transistor and why is it used?

3.16. What is meant by the terms fan-in and fan-out when applied to an electronic gate?

3.17. List the relative advantages of t.t.l. and low-power Schottky logic elements.

3.18. What is the result of exceeding the fan-out of an electronic gate?

3.19. What are the relationships between (i) the gate resistor and the gate load, (ii) the gate resistor and the current taken by the inputs, and (iii) the gate load and the fan-out in a diode gate?

3.20. Explain what is meant by the storage delay of a switching transistor and give one method by which it can be reduced.

3.21. State the meanings of the terms (i) threshold level, (ii) logic level as applied to an electronic gate.

3.22. Explain what is meant by the noise immunity of an electronic gate.

3.23. Describe a method of removing a digital integrated circuit from a printed circuit board.

3.24. What are the major disadvantages of diode logic compared with t.t.l. logic?

3.25. Explain the operation of the circuit given in Fig. 3.14.

3.26. State the precautions needed when handling logic i.c.s.

4 Bistable Multivibrators

A **bistable multivibrator** is a circuit that has two stable conditions and is able to remain in either one for an indefinite length of time. The circuit will change state only when a switching operation is initiated by a trigger pulse applied to the appropriate terminal. Once switched the bistable or **flip-flop** will remain in its other stable state until another trigger pulse is received that will force it to revert to its original state. The flip-flop has two output terminals, normally labelled Q and $\bar{Q}$.

When it is in the state $Q = 1, \bar{Q} = 0$, it is said to be SET.

When it is in the state $Q = 0, \bar{Q} = 1$ the circuit is said to be RESET.

Four types of flip-flop are available, known as the S-R, the J-K, the D, and the T flip-flops, each having their own particular fields of application. Very often a flip-flop is CLOCKED or STROBED, that is, it is operated in synchronism with a pulse train derived from a free-running multivibrator or a crystal oscillator known as the CLOCK.

Bistable multivibrators can be designed using discrete components and bipolar or field effect transistors or operational amplifiers [EIII] but such circuits will not be considered here. In this book emphasis will be placed on the suitable interconnection of NAND or NOR gates to form the various kinds of flip-flop, and on the flip-flops which are available in the t.t.l. and the c.m.o.s. families of digital integrated circuits.

The S-R Bistable

The symbol for an S-R flip-flop is given in Fig. 4.1. The device has two input terminals labelled S and R and two output terminals labelled Q and $\bar{Q}$. Always the logical state of the $\bar{Q}$ output is the complement of the state of the Q terminal. The logical operation of an S-R flip-flop is summarized by its truth table, Table 4.1.

The symbol Q^+ represents the logical state of the Q output *after* a set (S) or a reset (R) pulse has been applied to the appropriate input terminal. When $S = R = 0$ the state of the output will remain unchanged at whatever logical state it should already have. To *set* the circuit, that is to make $Q = 1$ and $\bar{Q} = 0$, requires $S = 1, R = 0$. If the circuit was already set before the S pulse is applied, the flip-flop will not switch. Similarly, to reset the circuit, $Q = 0, \bar{Q} = 1$, requires $S = 0, R = 1$. If pulses are simultaneously applied to both the set and the reset terminals, so that $S = R = 1$, the effect upon the circuit

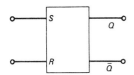

Fig. 4.1 Symbol for an S-R flip-flop

Table 4.1 S-R flip-flop

S	R	Q	Q^+
0	0	0	0
0	0	1	1
1	0	0	1
1	0	1	1
0	1	0	0
0	1	1	0
1	1	0	X
1	1	1	X

cannot be predicted; the flip-flop may switch to reverse the states of its two outputs or it may remain in its existing condition. The $S = R = 1$ condition is said to be *indeterminate*.

Very often the S-R flip-flop is known as a **latch**.

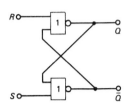

Fig. 4.2 S-R flip-flop using two NOR gates

1 Fig. 4.2 shows how an S-R flip-flop can be made by connecting together two NOR gates. The output of a 2-input NOR gate is 1 only when both of its inputs are at 0; if either or both of its inputs is at logical 1, the output state will be 0.

Suppose the circuit is initially set, i.e. $Q = 1, \bar{Q} = 0$. If $S = R = 0$ the upper gate will then have both of its inputs at 0 and so Q remains at 1. The lower gate has one input at 0 and the other at 1, hence its output $\bar{Q} = 0$. If now a pulse is applied to the input (reset) terminal only, giving $S = 0, R = 1$, the upper gate will have one input (R) at 1 and the other ($\bar{Q}$) at 0; the output (Q) of this gate will then become equal to logical 0. Both inputs to the lower gate are now at 0 and so the output of this gate, the $\bar{Q}$ terminal, becomes logical 1. Thus the circuit has switched states from $Q = 1, \bar{Q} = 0$ to $Q = 0, \bar{Q} = 1$, i.e. the flip-flop has been *reset*.

With the flip-flop in the reset condition, suppose that the input conditions change to $S = 1, R = 0$. The lower gate will now have one input (S) at 1 and the other (Q) at 0 and so its output ($\bar{Q}$) will be 0. This means that the upper gate now has both of its inputs at 0 and so its output (Q) becomes 1.

Whether the flip-flop is set or reset the application of a pulse to both the S and the R inputs may or may not switch the circuit; in other words the circuits operation is indeterminate.

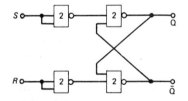

Fig. 4.3 S-R flip-flop using four NAND gates

2 The S-R flip-flop can also be constructed using four NAND gates interconnected in the manner shown in Fig. 4.3. The output of a NAND gate is 0 only if both of its input terminals are at 1; if either or both inputs is at 0, the output will be 1. Suppose that initially the flip-flop is set and $S = R = 0$. Both inputs to the lower gate are 1 and hence $\bar{Q}$ remains at 0. This means that the upper gate has one of its inputs ($\bar{S}$) at 1 and the ($\bar{Q}$) at 0 and so its output terminal remains at 1. Resetting the circuit requires the input condition $S = 0, R = 1$. Then, the lower gate will have one input ($\bar{R}$) at 0 and its other input (Q) at 1 and so its output ($\bar{Q}$) will become 1. This results in the upper gate having both its inputs ($\bar{S}$ and $\bar{Q}$) at 1 and its output becomes $Q = 0$. To set the circuit $S = 1, R = 0$; then the upper gate has one input ($\bar{S}$) at 0 and the other ($\bar{Q}$) at 1 and hence its output is $Q = 1$. Finally, the lower gate has now both of its inputs at 1 and it switches to give $\bar{Q} = 0$. Once again the input state $S = R = 1$ is indeterminate and may result in the flip-flop either switching or remaining unchanged.

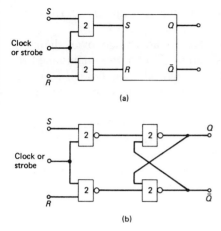

(a)

(b)

Fig. 4.4 Gated S-R flip-flop

3 Very often it is desirable for the set and reset operations to occur at particular instants in time determined by the *clock*. A *clocked* or *strobed* flip-flop will change only when a clock pulse is received. This operating characteristic is readily achieved by the use of suitable input gating. Fig. 4.4*a* shows one method that can be employed. The output of an AND gate is 1 only when both of its inputs are at 1. This means that when $S = 1, R = 0$ the flip-flop will not be set until the clock also becomes logical 1, since until this moment the S terminal of the flip-flop is not 1. Similarly, when $S = 0$, $R = 1$ the circuit will not reset until a clock pulse is present. The truth table of a clocked, or gated, S-R flip-flop is given in Table 4.2.

Table 4.2 Clocked S-R flip-flop

S	0	0	1	1	0	0	1	1	0	0	1	1	0	0	1	1
R	0	0	0	0	1	1	1	1	0	0	0	0	1	1	1	1
Clock	0	0	0	0	0	0	0	0	1	1	1	1	1	1	1	1
Q	0	1	0	1	0	1	0	1	0	1	0	1	0	1	0	1
Q^+	0	1	0	1	0	1	0	1	0	1	1	1	0	0	X	X

An alternative method of clocking a NAND gate type of S-R flip-flop is given by Fig. 4.4*b*. Whenever the clock is 0, the outputs of both gates must be 1 whatever the logical states of the S and the R inputs. Suppose the flip-flop is set; then $Q = 1, \bar{Q} = 0$ and so the upper right-hand gate has one input at 1 $(\bar{C}S + \bar{C}\bar{S})$ and one at 0 $(\bar{Q})$ and hence the Q output remains at 1. The lower right-hand gate has both of its inputs at logical 1 and so the $\bar{Q}$ output remains unchanged at 0. Only when the clock input is 1 can the appropriate input gate (S or $R = 1$) have an output at 0 for a possible switching action to be initiated. In other words, when the clock or strobe input is at logical 1, the circuit follows the same sequence of operations as the basic NAND flip-flop of Fig. 4.3. The strobe determines the times at which the S and R input signals should be effective and this is illustrated by the waveforms given in Fig. 4.5. It can be seen that the flip-flop does not immediately set

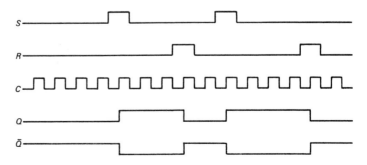

Fig. 4.5 Waveforms in a gated S-R flip-flop

once an S pulse is received but the switching is delayed until the clock is also 1.

The strobed S-R latch has an advantage over the non-strobed version in that its output state can be maintained (or *latched*) for any required time period even though S and R pulses are received by suitable choice of the clock frequency.

The S-R flip-flop, or latch, is available in both the t.t.l. and the c.m.o.s. families and examples are given in Appendices 1 and 2. As an example the pin connections of the 4044 c.m.o.s. circuit are given in Fig. 4.6; this is a quad NOR S-R latch.

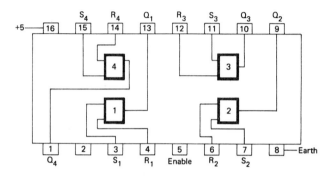

Fig. 4.6 Pin connections of a 4044 quad NOR S-R latch

In the explanations given of the operation of the S-R flip-flop it has been assumed that the inputs S and R change at the same time to initiate a change in the output state. Thus if the circuit is set and is to be reset, the required input condition is $S = 0, R = 1$. If the circuit has just been set with $S = 1$ and $R = 0$, both of the inputs must change; S from 1 to 0 and R from 0 to 1. Correct operation will be obtained if these changes take place simultaneously. However, should the R input change before the S input changes from 1 to 0, the indeterminate condition exists and the circuit may or may not switch. It will, of course, switch once the S change has occurred but an unwanted time delay may have been introduced. Similarly, if the S input should change from 1 to 0 before the R input changes from 0 to 1, then for some short time both inputs are at 0 and the circuit will not switch; again a time delay is introduced.

Clocked or strobed operation of an S-R flip-flop, and of the other types of flip-flop discussed later in this chapter, ensures that all the flip-flops in a system operate in synchronism with one another. Synchronous operation of a digital system is generally advantageous since it

(*a*) leads to faster operation,
(*b*) avoids the transient problems which may arise in a non-synchronous system.

Table 4.3 J-K flip-flop

J	K	Q	Q⁺
0	0	0	0
0	0	1	1
1	0	0	1
1	0	1	1
0	1	0	0
0	1	1	0
1	1	0	1
1	1	1	0

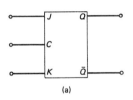

(a)

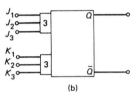

(b)

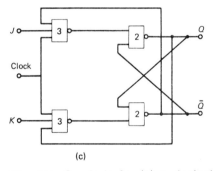

(c)

Fig. 4.7 Symbols for (a) a clocked J-K flip-flop, (b) a gated J-K flip-flop (c) NAND gate J-K flip-flop

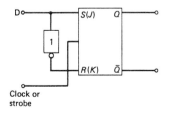

Fig. 4.8 D flip-flop

The J-K Flip-flop

For many digital applications the indeterminate $S = R = 1$ state of an S-R flip-flop cannot be permitted and, when this is the case, an alternative circuit, known as the J-K flip-flop, can be used. The operational difference between the S-R and the J-K flip-flops lies in the final two rows of their truth tables. This can be seen by comparing the truth table of the J-K flip-flop (Table 4.3) with Table 4.1.

As with the S-R flip-flop the symbol Q^+ represents the state of the Q output after a J or K trigger pulse has been applied to the circuit. The J pulse acts as the set signal, and the K pulse acts as the reset signal. The last two rows of Table 4.3 show that the J-K flip-flop *always* changes state when *both* J and K pulses are simultaneously applied to the circuit.

The symbol for a J-K flip-flop is given in Fig. 4.7a. A clock input is shown since this type of flip-flop is normally operated synchronously. Very often, clear terminals and perhaps reset terminals are also provided. Some J-K flip-flops are *gated,* that is they are provided with AND gates inside the integrated circuit package. The AND gate symbols are drawn touching the flip-flop symbol to indicate that the gates are internally provided (Fig. 4.7b). The J input of the flip-flop will be at logical 1 only when all three inputs J_1, J_2 and J_3 are at 1. Similarly, the K input is 1 only when $K_1 = K_2 = K_3 = 1$.

A J-K flip-flop can be constructed by ANDing the S input of an S-R flip-flop with its $\bar{Q}$ output and ANDing the R input with its Q output. The NAND gate version is shown in Fig. 4.7c and this circuit should be compared with Fig. 4.4b. The operation of this circuit is left as an exercise (Exercise 4.11).

The D Flip-flop

A D flip-flop has a single trigger (D) input terminal and its logical operation is such that its Q output terminal *always* takes up the same logic value as the D input. Any change in state takes place when the clock or strobe is 1. The D flip-flop is easily derived from an S-R flip-flop or a J-K flip-flop by connecting an inverting stage between the S and the R, or between the J and the K terminals as shown by Fig. 4.8. This connection means that always $R = \bar{S}$, or $K = \bar{J}$.

Substituting in the truth table of a J-K flip-flop (or an S-R flip-flop) gives Table 4.4, since when $J = 0$, $K = 1$, and when $J = 1$, $K = 0$.

It is clear from Table 4.4 that the Q output of a D flip-flop is always equal to the logical value of its D ($= J = S$) input. The D flip-flop is always strobed and so changes in its output state can only occur when the strobe input is 1. When the strobe or

Table 4.4 D flip-flop

J	K	Q	Q⁺
0	1	0	0
1	0	0	1
0	1	1	0
1	0	1	1

Table 4.5 Clocked D flip-flop

Cl	D	Q	Q⁺
1	0	0	0
1	0	1	0
1	1	0	1
1	1	1	1

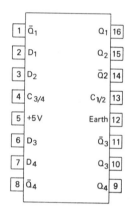

Fig. 4.9 Pin connections of 7475 quad D latch

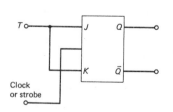

Fig. 4.11 T flip-flop

clock is 0, the output cannot follow the D input, i.e. the output is *latched* The truth table of a clocked D flip-flop is given by Table 4.5.

Integrated circuit versions of the D flip-flop are available in both the t.t.l. and the c.m.o.s. families; for example the 7475 t.t.l. dual D flip-flop and the 4042 c.m.o.s. quad D latch. It should be noticed that reference has been made to both the D flip-flop and the D latch. The operational difference between the two is that while the output of the flip-flop can only change state when the clock pulse changes from 1 to 0 (or, for some types, from 0 to 1), the latch is able to change its output state at any time while the clock is at 1. Fig. 4.9 shows the pin connections of the 7475 quad D latch.

If a D flip-flop has its $\bar{Q}$ output connected to its D input, the circuit will divide by two the signal applied to its clock input. The necessary connection is shown in Fig. 4.10a. Suppose that initially $Q = 1$, $\bar{Q} = D = 0$, then the first clock 1 pulse will cause the circuit to switch to $Q = 0$, $\bar{Q} = D = 1$. The second clock pulse will now switch the circuit back to $Q = 1$, $\bar{Q} = D = 0$ and so on. The circuit waveforms are given by Fig. 4.10b from which it is clear that the output pulse waveform has a frequency of one-half that of the input waveform.

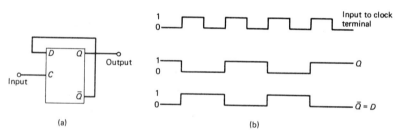

Fig. 4.10 Use of a D flip-flop to divide by two

The T Flip-flop

The fourth type of flip-flop is called the trigger or T flip-flop and its symbol is given by Fig. 4.11. The flip-flop is made from a J-K flip-flop by merely connecting its J and K terminals together. This means that, at all times, $J = K$. Substituting in the truth table of a J-K flip-flop gives Table 4.6. From this table it is apparent that, provided the clock or strobe is 1, the flip-flop will change state, or **toggle**, each time there is a trigger (T) pulse applied to its input. Thus, when $T = 0$, the Q output will not change state when the clock goes to logical 1; when $T = 1$, the Q output will change state each time a clock pulse is received (Fig. 4.12). The T flip-flop is not available as an integrated circuit since it is so easily obtained from a J-K flip-flop. Its truth table is given in Table 4.7.

Fig. 4.12 Waveforms in a T flip-flop

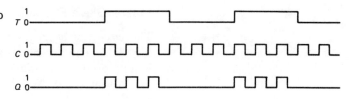

Table 4.6

J	K	Q	Q⁺
0	0	0	0
0	0	1	1
1	1	0	1
1	1	1	0

Table 4.7 T flip-flop

Cl.	T	Q	Q⁺
1	0	0	0
1	0	1	1
1	1	0	1
1	1	1	0

Master-Slave and Edge-triggered Flip-flops

When a flip-flop is synchronously operated, i.e. enabled by a clock or strobe pulse, the circuit must respond to any changes in the input signal that take place *during* the clock 1 period. This means that the duration of a clock pulse must be less than the time it takes for the flip-flop to change state in response to an input trigger signal. This often leads to practical difficulties, such as false switching, arising from the rise-times and fall-times of the clock pulses being insufficiently short. Thus, referring to Fig. 4.13, a J pulse occurring at time t_1 will probably set the flip-flop because at this instant in time the clock pulse has reached about 50% of its final voltage, but a J pulse arriving at time T_2 would not trigger the circuit because

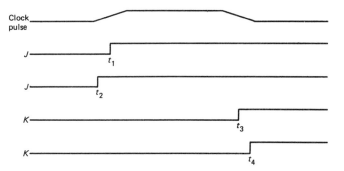

Fig. 4.13 Clock pulse rise-times and fall-times

although a clock pulse has arrived its amplitude is still very small. Similarly, a K pulse initiated at time t_3 might well reset the circuit even though the clock pulse has ended because the fall-time of the clock pulse voltage is such that its value is still fairly large. A K pulse at time t_4, on the other hand, would not reset the circuit since the clock pulse voltage has by now decreased to very nearly zero.

The J-K flip-flop circuit of Fig. 4.7c suffers from the possibility of a *race condition* occurring when $J = K = 1$. This condition results in the Q output oscillating between 1 and 0 while the clock is at 1, and then when the clock goes to 0, staying in either state. Since this state is unpredictable, the race condition cannot be tolerated.

The switching difficulties associated with clocked operation of a flip-flop can be overcome by the use of either *master-slave* or *edge-triggered* operation.

Master-Slave Flip-flop

The circuit of a master-slave bistable or flip-flop is shown in Fig. 4.14. The left-hand flip-flop is the *master* and the right-hand flip-flop is the *slave*. Suppose that initially the flip-flop is set so that $Q = 1$. When the clock is 0, the master will have $J = K = 0$ and hence its outputs Q' and $\bar{Q}'$ remain at 1 and 0 respectively. The clock input to the slave is inverted and so the slave has $J = 1, K = 0$. This condition sets the slave to retain Q and $\bar{Q}$ at 1 and 0 respectively.

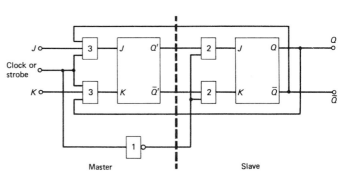

Fig. 4.14 Master-slave J-K flip-flop

When a clock pulse arrives, $C = 1$, either J or K or both could be at 1. If $J = 1, K = 0$ both inputs to the master are at 0 (since $\bar{Q} = 0$) and switching is not initiated. Suppose that now $J = 0, K = 1$; the three inputs to the lower input AND gate are all at 1 and so the master flip-flop has inputs $J = 0, K = 1$ applied. The master therefore changes state to have $Q' = 0, \bar{Q}' = 1$. However, since the clock is inverted, this change of state cannot be passed onto the slave, both of whose inputs remain at 0, so that at the output the condition $Q = 1, \bar{Q} = 0$ is retained. At the end of the clock pulse $C = 0$ and now the lower slave input AND gate has both its inputs at 1. Therefore, the slave has $J = 0, K = 1$ and switches states to give $Q = 0, \bar{Q} = 1$. Thus the slave is reset by the trailing edge of the clock pulse. Lastly, consider that $J = K = 1$. Since the circuit is set $\bar{Q} = 0$ and one of the three inputs to the upper input AND gate is 0 and hence the J input of the master is also 0. The lower input gate now has all three inputs at 1 when a clock pulse arrives and so the master has $K = 1$. This means that the operation when $J = K = 1$ is the same as for the condition $J = 0, K = 1$ just described.

Consider now the circuit operation when the flip-flop is initially reset, i.e. $Q = 0, \bar{Q} = 1$. When $J = 1, K = 0$, the J input of the master will become 1 once a clock pulse is present but the K input will be at 0. Hence $Q' = 1, \bar{Q}' = 0$ but state of Q' cannot be passed onto the slave's J input because the inverted clock pulse inhibits the slave input AND gates. When the

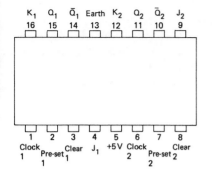

Fig. 4.15 Pin connections of 7476 dual J-K master-slave flip-flop

clock changes from 1 to 0, the J input of the slave will become logical 1 and the slave will be set to have $Q = 1$, $\bar{Q} = 0$. The input state $J = 0$, $K = 1$ will not initiate switching since for this condition neither gate will have all its three inputs at 1.

Master-slave J-K flip-flops are readily available in the t.t.l. and c.m.o.s. families and Fig. 4.15 shows the pin connections of the 7476, a t.t.l. circuit providing dual J-K master-slave flip-flops with pre-set and clear. *Pre-set* and *clear* terminals are provided with this particular integrated circuit and act like set and reset terminals which can be used to control the output of the flip-flop independently from the J, K and clock inputs.

When the preset and the clear inputs are at logic 1, the clock at 1 will enable the J and K inputs. The truth table describing the logical operation is given by Table 4.8.

In the table the symbol X denotes that the logic level is irrelevant. Other J-K master-slave flip-flops, for example the 7473, are not provided with a preset terminal. The truth table is then similar to Table 4.8 except that the top two rows and, of course, the preset column are omitted.

Table 4.8 Master-slave J-K flip-flop

Preset	Clear	Clock	J	K	Q	Q^+
0	1	X	X	X	0	1
0	1	X	X	X	1	1
1	0	X	X	X	0	0
1	0	X	X	X	1	0
1	1	trailing edge 1—0	0	0	0	0
1	1	trailing edge 1—0	0	0	1	1
1	1	trailing edge 1—0	1	0	0	1
1	1	trailing edge 1—0	1	0	1	1
1	1	trailing edge 1—0	0	1	0	0
1	1	trailing edge 1—0	0	1	1	0
1	1	trailing edge 1—0	1	1	1	0
1	1	trailing edge 1—0	1	1	0	1
1	1	1	X	X	0	0
1	1	1	X	X	1	1

The S-R flip-flop can also be manufactured in a master-slave version, the circuit being similar to that shown in Fig. 4.14, differing from it in that the connections between the Q and the $\bar{Q}$ outputs and the AND gates are not used. The operation of the S-R master-slave flip-flop is left as an exercise (4.3).

Edge-triggered Flip-flops

As an alternative to the use of the master-slave principle a clocked flip-flop can be *edge-triggered*. With a flip-flop of this type, changes in the output state are initiated by *changes* in the clock pulses. Any change in the input signal (J, K or D) that should occur while the clock is steady at logical 1 will not affect the output of the circuit. In general, standard t.t.l. flip-flops are master-slave circuits, whilst the low-power Schottky versions are edge-triggered.

Edge-triggered flip-flops, both J-K and D, are available in the t.t.l. family, e.g. 7470 is a J-K flip-flop circuit and 7474 is a dual D flip-flop.

Uses of Bistable Multivibrators

All types of flip-flop have many applications in the control of events taking place in digital circuits where time delay or bit storage is required. S-R flip-flops are often used as switch contact de-bouncing circuits. J-K and D flip-flops can be used as divide-by-2 circuits and as stages in counters and in shift registers.

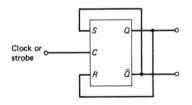

Fig. 4.16

Exercises

4.1. Draw the circuit diagram of a clocked S-R flip-flop using NAND gates only and explain its action. Write down the truth table of the circuit.

4.2. Explain, with the aid of a truth table, the operation of the circuit shown in Fig. 4.16.

4.3. Draw the circuit diagram of an S-R master-slave bistable and describe its operation. Is its operation indeterminate for the input condition $S = R = 1$?

4.4. (*a*) With the aid of a circuit diagram, explain the operation of a bistable device which has gated inputs for typical use in a register.

(*b*) Explain briefly the meanings of the terms master and slave when applied to a clocked bistable device. (*C & G*)

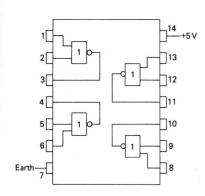

Fig. 4.17

4.5. (*a*) Show how an S-R flip-flop can be constructed using NAND gates only.

(*b*) Modify the circuit to act as a D flip-flop.

(*c*) Show how the D flip-flop circuit can be used as a clocked divide-by-two circuit.

4.6. Fig. 4.17 shows the pin connections of a 4001 c.m.o.s. quad 2-input NOR gate. Explain, with the aid of diagrams, how the i.c. could be connected to form a divide-by-two circuit. Discuss the precautions needed when handling c.m.o.s. circuits.

4.7. Explain, with the aid of sketches, the difference between synchronous and non-synchronous operation of a J-K flip-flop. Why is synchronous operation sometimes used? Write down the truth table of synchronous and non-synchronous J-K flip-flops.

4.8. (*a*) With the aid of a logic diagram, show how two NOR gates may be interconnected to form a clock-controlled bistable multivibrator (R-S flip-flop). Give a truth table showing all possible changes of state. With the aid of waveforms, describe a typical sequence of operations.

(*b*) Explain why certain changes of state of the bistable in (*a*) are indeterminate, and with the aid of a logic diagram, explain how this problem is avoided in a J-K flip-flop.   (*C & G*)

4.9. Fig. 3.12 *a* shows the pin connections of a 7400 quad 2-input NAND gate. Show how it can be used to produce (i) an S-R flip-flop, (ii) a D flip-flop. Can a T flip-flop be constructed? Justify your answer using a truth table.

4.10. Explain, with the aid of a truth table, the logical operation of a master-slave flip-flop. Could the circuit be modified to act as (i) a D flip-flop, (ii) a T flip-flop?

Short Exercises

4.11. Explain the operation of the J-K flip-flop circuit of Fig. 4.7*c*.

4.12. Write down the truth table of a J-K flip-flop. Explain how it differs from the truth table of an S-R flip-flop.

4.13. Explain why a J-K flip-flop can be used as a divide-by-two device.

4.14. Show how a D flip-flop can be fabricated using NOR gates only.

4.15. What is the difference between setting or resetting and toggling a flip-flop?

4.16. A J-K flip-flop is in the state $Q = 1$, $\bar{Q} = 0$. What change in state will occur if the flip-flop is (*a*) set, (*b*) reset, (*c*) cleared, (*d*) clocked?

4.17. Why cannot a T flip-flop be purchased in integrated form? How can a T flip-flop be obtained?

5 Counters

A **counter** is an electronic circuit that is able to count the number of pulses applied to its input terminals. The count may be outputted using the straightforward binary code, or be in binary-coded decimal, or in some other version of binary notation. Alternatively, the outputs of a counter may be decoded to produce a unique output signal to represent each possible count. A counter may also be used as a frequency divider, in which case only one output terminal is required.

Essentially, a counter consists of the tandem connection of a number of flip-flops, usually of the J-K type which may be operated either synchronously or non-synchronously. With **synchronous operation** all the flip-flops making up the counter operate at the same instant in time under the control of a clock pulse. In the case of **non-synchronous operation** each flip-flop operates in turn. The switching of the least significant flip-flop is initiated by a clock pulse but the remaining flip-flops are each operated by the preceding flip-flop. This means that each stage must change state before the following stage can do so. As a result synchronous operation of a counter is much faster and the use of a non-synchronous counter is only acceptable when the speed of operation is not of particular importance. On the other hand a synchronous counter is more complex and so is more expensive.

A counter can be constructed by suitably interconnecting a number of integrated circuit J-K flip-flops and, perhaps, gates, but more conveniently several different types of counter are available in integrated circuit form and in this chapter examples will be given of the use of t.t.l. counters.

The possible applications for counters are many. They are often used for the direct counting of objects in industrial processes and of voltage pulses in digital circuits such as digital voltmeters. Counters can be used as frequency dividers and for the measurement of frequency and time.

Non-synchronous Counters

A single J-K flip-flop will act as a divide-by-two circuit. If its J and K terminals are both held at logical 1 and a clock or strobe pulse is applied to its clock input terminal, the flip-flop will toggle and so the Q output will switch backwards and forwards between logical 1 and logical 0 (Fig. 5.1). In the drawing of the waveforms shown in Fig. 5.1 it has been assumed that the flip-flop toggles at the trailing edge of the

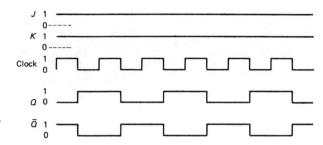

Fig. 5.1 Waveforms in a divide-by-two circuit

clock pulse, i.e. as the clock pulse changes from 1 to 0. This is generally true for non-synchronous counters. Clearly the number of pulses per second occurring at the Q terminal is only one-half of the number of clock pulses. For counts in excess of two, a number n of J-K flip-flops must be connected in cascade to give a maximum count of $2^n - 1$. Since the first count is 0 this means that n flip-flops are able to give a count of 2^n.

1 Fig. 5.2 shows how a non-synchronous counter can be constructed from four J-K flip-flops. The J and the K inputs of each flip-flop are permanently connected to the logical 1 voltage level so that each flip-flop will be toggled by a pulse

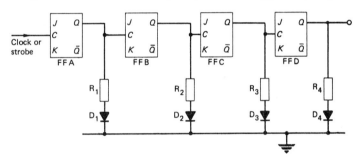

Fig. 5.2 Non-synchronous counter

applied to its clock input. The connections of the J and K terminals to 1 are not shown in the diagram but should be understood. The Q output of the first, second, and third stages is connected to the clock input of the following stage. The $\bar{Q}$ terminals are left unconnected. Each J-K flip-flop is switched by the trailing (falling) edge of a pulse applied to its clock input. In the figure the logical state of each stage is indicated by a **light-emitting diode** (l.e.d.) connected from the Q terminal to earth via a current-limiting resistor. An l.e.d. is a semiconductor device which emits visible light when it is forward biased and passes a sufficiently large current. Different versions of the device are available in which the colour of the visible light may be red, green or yellow. Current will flow through an l.e.d. and the device will visibly glow whenever the associated Q terminal is at logical 1 (+5 V).

Suppose that initially each stage is reset, i.e. $Q_A = Q_B = Q_C = Q_D = 0$. The trailing edge of the first clock pulse will toggle flip-flop A so that Q_A becomes logical 1. Only l.e.d. D_1 will emit visible light so that the displayed count will be 0 0 0 1 (reading from the right) or denary 1. The next clock pulse will toggle flip-flop A and the change of Q_A from 1 to 0 will set flip-flop B. Thus, after two clock pulses have been applied to the counter, only l.e.d. D_2 will glow visibly. The third clock pulse will set flip-flop A so that Q_A changes from 0 to 1 but such a change will not affect the state of flip-flop B and so this stage remains set. Now both the first two stages are set, $Q_A = Q_B = 1$, and the last two stages remain reset, $Q_C = Q_D = 0$; l.e.d.s D_1 and D_2 are lit and the displayed count is 0 0 1 1 or decimal 3. When the fourth clock pulse arrives, the first stage toggles so that Q_A changes from 1 to 0, and this change causes flip-flop B to reset. Thus Q_B changes from 1 to 0 and in so doing sets flip-flop C; now $Q_A = Q_B = Q_D = 0$ and $Q_C = 1$. This condition is indicated by only l.e.d. D_3 emitting visible light. The operation of the counter as the fifth, sixth, seventh, etc. clock pulses are applied follows the same lines as just described and can be summarized by the truth table of the counter (see Table 5.1).

Table 5.1 Counter

Clock pulse	0	1	2	3	4	5	6	7	8	9	10	11	12	13	14	15	16
Q_A	0	1	0	1	0	1	0	1	0	1	0	1	0	1	0	1	0
Q_B	0	0	1	1	0	0	1	1	0	0	1	1	0	0	1	1	0
Q_C	0	0	0	0	1	1	1	1	0	0	0	0	1	1	1	1	0
Q_D	0	0	0	0	0	0	0	0	1	1	1	1	1	1	1	1	0

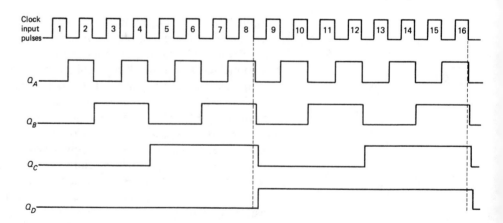

Fig. 5.3 Waveforms in a 4-bit non-synchronous counter

The operation of the counter can be illustrated by the waveforms given in Fig. 5.3. The Q outputs of the flip-flops seem to *ripple* through the circuit and for this reason this type of circuit is often known as a **ripple counter**.

The count, displayed by the l.e.d.s, is of a binary nature and the counter operates non-synchronously because the flip-flops operate at different times, shown at the end of the eighth and the sixteenth clock pulses by the vertical dotted lines. Flip-flop A operates twice as often as flip-flop B, four times as often as flip-flop C, and eight times as often as flip-flop D. There is a maximum clock frequency that can be used since the periodic time of the clock waveform must be greater than the sum of the propagation times through the counter and the time duration of the output pulse.

If the $\bar{Q}$ output of each flip-flop is connected to the clock input of the next flip-flop, the circuit will count down from 15 to 0.

Often a binary readout of the count is undesirable; when this is so the outputs of the individual flip-flops of the counter can be **decoded** so that each count produces a unique output. Decoding can be achieved by connecting the Q and the $\bar{Q}$ outputs of each flip-flop to the inputs of a number of gates.

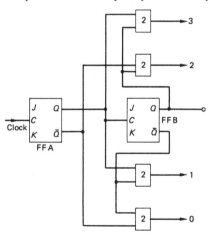

Fig. 5.4 Two-stage counter with decoded output

This is shown by Fig. 5.4 for a 2-stage counter. The output of a 2-input AND gate is 1 only when both its inputs are at 1. Hence the top gate, for example, will only have an output of logical 1 when $Q_A = Q_B = 1$ and the count is 3. Alternatively, decoding can be carried out using NOR gates (see example 5.2).

2 It is possible for false 1 signals to appear at the flip-flop outputs which may give false counts when decoding circuitry is used. These *glitches* or *dynamic* hazards arise because not all the flip-flops change state at precisely the same time when the

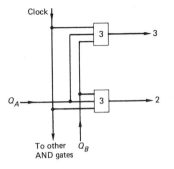

Fig. 5.5 Clocking of decoding gates in a counter

edge of a clock pulse arrives. To prevent this happening the decoding gates can be clocked as shown by Fig. 5.5. The enabling clock pulse is not applied until all the flip-flops have reached their steady (final) values. The glitch-free output is, however, obtained at the expense of a reduction in the speed of operation. Some integrated circuit counters include the decoding circuitry within the package. Often the count is to be displayed by a **7-segment display**; this kind of display is commonly employed and consists of a number of l.e.d.s arranged in seven segments as shown in Fig. 5.6a. Any number between 0 and 9 can be indicated by lighting the appropriate segments; this is shown by Fig. 5.6b. Clearly the 7-segment display requires a 7-bit input signal and so a decoder is needed to convert the output of the counter into a 7-segment signal. Fig. 5.7 shows one arrangement, in which the b.c.d. output of a decade counter is converted to a 7-segment signal by a decoder. Usually the decoder first converts the b.c.d. signal into decimal and then converts the decimal number into 7-segment.

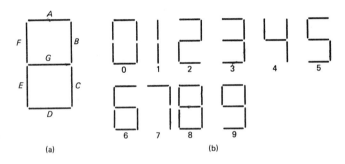

Fig. 5.6 7-segment display (a) arrangement of l.e.d.s, (b) indication of numbers 0 to 9

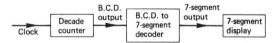

Fig. 5.7 B.C.D. decade counter with 7-segment display

When a count in excess of 9 is required, a second counter must be used and be connected in the manner shown by Fig. 5.8. The tens counter is connected to the output of the final flip-flop of the units counter in the same way as the flip-flops inside the counters are connected.

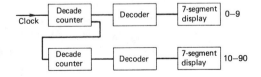

Fig. 5.8 Decade counters arranged to give a count in excess of 9

Reducing the Count to Less than 2^n

Very often a counter is required to have a count of less than 2^n, where n is the number of flip-flops it contains. For example the decade counter shown in Fig. 5.9 has a count of 10 obtained from a four flip-flop counter. The reduced count is obtained by modifying the basic counter circuit so that one or more of the possible counts are omitted. Thus, if a count of 7 is required, a three-stage counter must be used, having a maximum count of 2^3 or 8 (including 0). This means that *one* of the counts must be eliminated. There are three different ways in which one or more counts can be eliminated: (*a*) the feedback method, (*b*) the reset method, and (*c*) the preset method.

The Feedback Method

The feedback method of reducing the count of a counter consists of feeding back the output(s) Q and/or $\bar{Q}$ of one or more flip-flops to one or more of the preceding flip-flops in order to set or reset them out of their normal 2^n count sequence. The feedback method is easily applied to counters made up of separate J-K flip-flops (themselves integrated) but cannot be applied to an integrated circuit counter because these devices do not have their internal flip-flop J and K terminals available at the package pins.

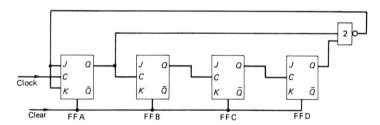

Fig. 5.9 Decade counter

Fig. 5.9 shows how a 4-stage counter can have feedback applied to reduce its count from 16 to 10. The output of a 2-input NAND gate is 0 only when both its inputs are at 1. Hence the NAND gate shown will have an output of 0 only when $Q_A = Q_D = 1$, i.e. when the count is 1001 or decimal 9. Then the J and the K inputs of flip-flop A are both at logical 0 and the output Q_A will not change as further clock pulses are received. The counter will therefore count up to 9 and then stop. The truth table of the counter is shown by Table 5.2. A clear pulse must be applied to reset the counter before another count can be commenced.

Table 5.2

Clock pulse	0	1	2	3	4	5	6	7	8	9	10
Q_A	0	1	0	1	0	1	0	1	0	1	0
Q_B	0	0	1	1	0	0	1	1	0	0	0
Q_C	0	0	0	0	1	1	1	1	0	0	0
Q_D	0	0	0	0	0	0	0	0	1	1	0

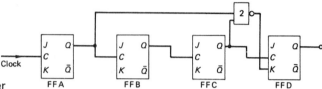

Fig. 5.10 An alternative decade counter

An alternative arrangement, which is really feed-forward, is shown in Fig. 5.10. The output of the NAND gate will be at 0 only when the flip-flops A and C are both set, i.e. when the count is 5. At this count the K input of flip-flop D is 0 and so this stage sets; now $Q_A = Q_C = Q_D = 1$ and so the count has jumped to 13. Further clock pulses now produce counts of 14 and 15 before the counter restores to its original state with zero count.

The Reset Method

Many counters, discrete or integrated, can have their 2^n count reduced by resetting the circuit at the appropriate point in the count. The only requirement is that some means of resetting, or clearing, all the flip-flops simultaneously should be provided. To construct a counter with a count of N, the number n of flip-flops needed is such that $2^n > N$, i.e. if $N = 5$, 6 or 7, three flip-flops will be needed since $2^n = 8$. When the count of the unmodified counter is N, the Q output of each flip-flop that is then at binary 1 must be connected to the input of a NAND gate. The output of this NAND gate should then be connected to the reset terminal of the counter; for i.c. counters this is normally activated by the logic 0 voltage level.

Fig. 5.11a shows a 3-stage ripple counter having a maximum count of $2^3 = 8$. Suppose that this counter is to be converted to have a count of 6. When the count is 6 then flip-flops B and C are set and hence their Q outputs are connected to a NAND gate and thence to the reset line (Fig. 5.11b). The output of the NAND gate will be logical 0 only when both of its inputs, i.e. Q_B and Q_C, are 1; this condition will occur at the end of the sixth clock pulse at which time the counter is reset. Hence the circuit will count through the sequence 0, 1, 2, 3, 4, 5, 0, 1, 2, etc.

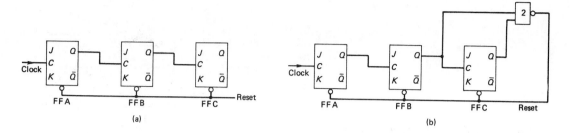

Fig. 5.11 (*a*) 3-stage counter, (*b*) divide-by-6 circuit

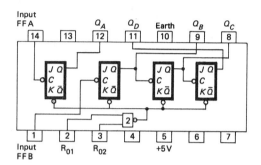

Fig. 5.12 7493 4-bit binary counter

As an example consider the t.t.l. 7493 4-bit binary counter shown in Fig. 5.12. It can be seen that flip-flops B, C and D are internally connected together to form a 3-stage ripple counter, but flip-flop A is not internally connected to the other three flip-flops. The terminals marked R_{01} and R_{02} are the "reset-to-0" inputs. A logical 1 voltage level applied to both of these terminals produces a logical 0 voltage level at the output of the associated NAND gate and this is the state required to reset, or clear, all four flip-flops (indicated by the small circle at the clear terminal of each flip-flop). For this integrated circuit to be used as a 0–15 counter, or a *divide-by-16* circuit, pin 12 should be connected to pin 1 to give a 4-stage ripple counter.

The 7493 can be made to have a count, or a division ratio, of less than 16 by use of the R_{01} and R_{02} input terminals. To obtain a **decade counter** pin 9 should be connected to pin 2; pin 11 should be connected to pin 3; and pin 12 should be connected to pin 1. The circuit is then set up as shown in Fig. 5.13. When the count reaches 10, $Q_B = Q_D = 1$ and the output of the NAND gate becomes 0, resetting all four flip-flops.

The arrangement shown in Fig. 5.12 is not the only way in which a non-synchronous counter can be made. For example, the t.t.l. 7490 is a decade counter that consists of one J-K flip-flop which acts as a *divide-by-two* circuit. This circuit can be externally connected to a 3-stage counter that is internally

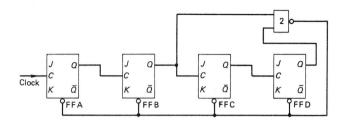

Fig. 5.13 7493 connected as a decade counter

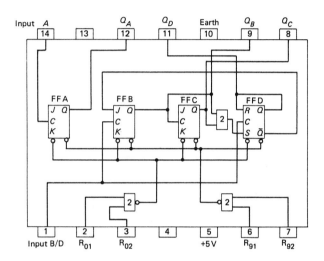

Fig. 5.14 Pin connections of a 7490 counter

connected to give a count of 5. The overall count is thus $2 \times 5 = 10$. In addition to the clear or reset terminals marked R_{01} and R_{02}, this particular integrated circuit has two other pins, labelled as R_{91} and R_{92}, which can be used to set the counter to a count of 9. The pin connections of the 7490 are shown in Fig. 5.14.

The Preset Method

Some counters are provided with a preset input terminal. A signal applied to this input will set all the flip-flops to have $Q = 1$. The method of modifying a preset counter to have a reduced count is similar to (but not the same as) that used in conjunction with a reset counter. Suppose a count of N is wanted where $N = 2^n$ ($n =$ number of flip-flops as before). Each flip-flop whose Q output is 1 when the count of the unmodified counter is N should have this output connected to one of the inputs of a NAND gate. Also applied to an input of the NAND gate should be the clock. The output of the NAND gate is taken to the preset terminal of each of the flip-flops then in the state $Q = 0$. Consider once again the 3-stage ripple

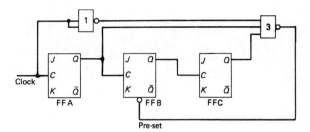

Fig. 5.15 Divide-by-5 circuit

counter given in Fig. 5.11a and suppose that it is to be converted to have a count of 5. When the count reaches 5, flip-flops A and C are set; hence a 3-input NAND gate is required whose inputs are connected to the clock, to Q_A, and to Q_C. The output of the NAND gate is connected to the preset terminal of flip-flop B only, as shown by Fig. 5.15.

The counter counts on the normal sequence until the end of the fifth clock pulse. At this point $Q_A = Q_C = 1$ and the output of the NAND gate will become logical 0 since the clock input is inverted. Flip-flop B is then set so that now $Q_A = Q_B = Q_C$. The trailing edge of the next clock pulse will now reset all the flip-flops to $Q = 0$ for a new count to start.

Clocks

Most digital circuits use some form of rectangular pulse generator, known as the *clock*, to control the times at which the various stages change state. At lower frequencies the clock may consist of an astable multivibrator [EIII] or a Schmidt trigger oscillator but at higher frequencies the clock is usually some form of crystal oscillator in order to achieve good frequency stability.

1 A **Schmidtt trigger** is a circuit, readily available as an integrated circuit, whose output voltage can have only one of two possible values. The output voltage will be high, +3.3 V for a t.t.l. version, when the input is greater than a positive-going *threshold voltage,* and will remain at this value until such time as the input voltage falls below the negative-going threshold voltage. The operation of a Schmidt trigger circuit is shown by Fig. 5.16a; when the input sinusoidal voltage becomes more positive than the positive threshold voltage, the output of the circuit switches to become +3.3 V. The output voltage stays at +3.3 V until the input voltage falls below the negative threshold voltage and then the output voltage suddenly switches to very nearly 0 V. The output voltage will now stay at 0 V until the input again becomes more positive than the positive threshold voltage.

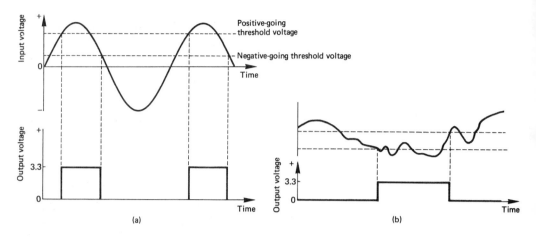

Fig. 5.16 (*a*) Operation of a Schmidtt trigger, (*b*) Operation of a NAND Schmidtt trigger

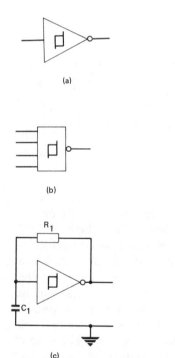

Fig. 5.17 Symbols for (*a*) Schmidtt trigger, (*b*) NAND Schmidtt trigger, (*c*) use of a Schmidtt trigger as a clock oscillator

The Schmidtt trigger is used to convert signals of varying waveshapes into rectangular pulses of short rise time and fall time. Some i.c. Schmidtt triggers are associated with 2-input or 4-input NAND gates; the gates allow the trigger to be enabled by the signals applied to the gates. Essentially a NAND Schmidtt trigger consists of a NAND gate followed by a Schmidtt trigger. The input to the trigger will be at a high voltage only when all the inputs to the NAND gate are low. This means that the operation is the inverse of that of the basic trigger. Suppose for example that the waveform shown in Fig. 5.16*b* is applied to the commoned inputs of a 2-input or a 4-input NAND Schmidtt trigger. The output of the circuit will be at zero volts until the input voltage first reaches the lower threshold voltage. When this point is reached, all the NAND inputs are low and so its output is high, causing the trigger output to rise to +3.3 V. The output will remain at +3.3 V until the input voltage rises to some value more positive than the input upper threshold voltage, and then it will abruptly switch to zero volts.

The symbols for a Schmidtt trigger and a NAND Schmidtt trigger are shown in Figs. 5.17*a* and *b* respectively. Also shown in *c* is a diagram of how a Schmidtt trigger can be used as a clock oscillator. Note that the Schmidtt trigger symbol indicates that its output is inverted, i.e. the output is low even the input is high. This is true for the t.t.l. version of the circuit—the 7414.

Fig. 5.18 Crystal oscillator clock

2 When a **crystal oscillator** is used as the clock, it is necessary to convert its sinusoidal output voltage into the required rectangular waveform; this is easily achieved with the use of a Schmidtt trigger as shown by Fig. 5.18.

3 Another way of producing a clock is by the suitable inter-connection of **logic gates**. Since both NAND and NOR gates include amplification, they can be used to form an oscillatory circuit because essentially an oscillator is merely an amplifier that provides its own input signal [EII and EIII]. One way in which NOR or NAND gates can be connected to form a clock is shown in Fig. 5.19a. The feedback network is provided by capacitor C and resistor R together with NOR gate B. The amplifier is provided by the NOR gate A. Suppose that initially the capacitor C is discharged so that the clock output is zero. The gate B inverts its input voltage and so its output voltage is at the logical 1 level. Capacitor C is therefore charged via R with a time constant of CR seconds. As C charges, the voltage across it rises until it becomes equal to the value corresponding to logic 1. Immediately the output of gate B goes to logical 0 and so the output of gate A—which is the required clock output—goes to logical 1. Now capacitor C commences to discharge via R to the output of gate B. Once C has discharged to the level at which the output of gate B switches back to logical 1, the output of gate A goes to logical 0. The frequency of operation of the clock is determined by the time constant CR.

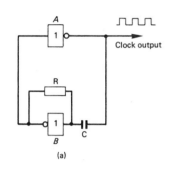

(a)

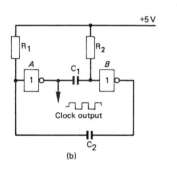

(b)

Fig. 5.19 Logic gate clock oscillators

An alternative method of making a clock from NOR or NAND gates connects them in an astable multivibrator arrangement, a possible circuit being given by Fig. 5.19b. The operation of astable multivibrators has been covered elsewhere [EIII].

Synchronous Counters

A ripple counter can only be used for applications in which the speed of operation is not very important. If several stages of counting are employed, the time taken for a clock pulse to *ripple* through the counter may well be excessive. The operating time can be shortened considerably by arranging for all the flip-flops to be clocked at the same moment. This is known as *synchronous* operation.

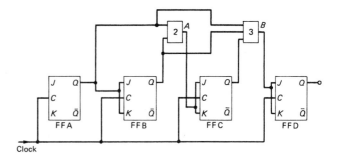

Fig. 5.20 Synchronous counter

Clock

In a synchronous counter all the flip-flops change their state simultaneously, the operation of each stage being initiated by the clock. The arrangement of a 4-bit synchronous counter is shown in Fig. 5.20. The clock input of each flip-flop is directly connected to the clock line so that they all operate simultaneously. The Q output of the first stage is connected to the commoned J and K inputs of the second stage, but the third and the fourth stages have their input state determined by the Q outputs of *all* the previous stages.

The operation of the *divide-by*-16 synchronous counter is as follows. Suppose that initially all the four stages are reset so that the count is 0, i.e. $Q_A = Q_B = Q_C = Q_D = 0$. At the trailing edge of the first clock pulse, flip-flop A toggles so that $Q_A = 1$. The second stage now has $J_B = K_B = 1$ and so it will toggle when the second clock pulse ends and Q_A changes from 1 to 0. Now $Q_A = 0$ and $Q_B = 1$. AND gate A now has one input at 1 and the other input at 0 and so flip-flop C has $J_C = K_C = 0$ and will *not* toggle at the end of the next clock pulse. When the third clock pulse ends, flip-flop A toggles, Q_A changes from 0 to 1, and flip-flop B remains set. Now $Q_A = Q_B = 1$, so the count is 3, and both inputs to gate A are at logical 1. This means that flip-flop C has $J_C = K_C = 1$ and will toggle when the fourth clock pulse ends. Thus, the fourth clock pulse resets flip-flops A and B and sets flip-flop C; now only $Q_C = 1$ and the count is 4. The operation of the counter continues in this way as further clock pulses are applied. All three inputs to AND gate B will be at 1 after the *seventh* clock pulse has ended and $Q_A = Q_B = Q_C = 1$. Thus flip-flop D will toggle to be set at the trailing edge of the eighth pulse.

The operation of the counter as 16 clock pulses are applied is summarized by the waveform diagram given in Fig. 5.21. In this figure the vertical dotted lines indicate the time delay that occurs between the clock pulse changing from 1 to 0 and the flip-flops toggling. Note that there is one output (Q_D) pulse for 16 input (clock) pulses.

The synchronous counter is faster to operate than the non-synchronous counter because the clock frequency is only lim-

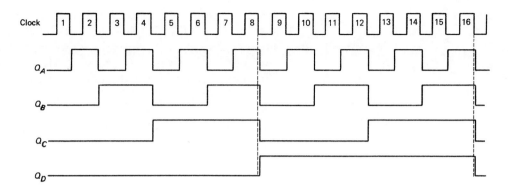

Fig. 5.21 Waveforms in a 4-bit synchronous counter

ited by the delay of *one* flip-flop (since all flip-flops operate simultaneously) plus the delays introduced by the AND gates. Another advantage is that a decoded synchronous counter does not usually suffer from dynamic errors or glitches.

Many integrated circuit synchronous counters operate at the rising edges of the clock pulses and not at the trailing edge as was shown by Fig. 5.21. This may sometimes be an advantage when a counter is interfaced with other digital circuitry. In some integrated counters the first and the second stages are not internally coupled together; the clock pulse is applied directly to the first stage and the remaining stages are simultaneously clocked by the output of the first stage. This kind of operation is often known as *semi-synchronous*.

Reducing the Count to Less than 2^n

The count of a synchronous counter can be reduced to less than 2^n using any one of the three methods introduced earlier in conjunction with ripple counters. An example of the feedback method is given in Fig. 5.22. Assuming the counter is initially cleared, the first clock pulse will toggle flip-flop A, $Q_A = 1$. The two inputs of AND gate A are now both 1 and so flip-flop B sets at the end of the second clock pulse. Now $Q_A = 0$, $Q_B = 1$. The J input of flip-flop B is now at logical 0 and

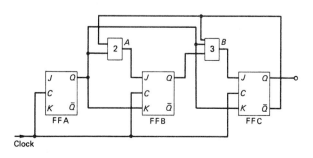

Fig. 5.22 Synchronous counter with feedback

hence the third clock pulse will only toggle flip-flop A to give the state $Q_A = Q_B = 1$. Now all of the inputs to AND gate B are 1 and so all three flip-flops toggle when the fourth clock pulse ends. $\bar{Q}_C$ is now 0 and so AND gates A and B are inhibited. The fifth pulse therefore toggles flip-flop A only to give $Q_A = Q_C = 1$. Flip-flop C now has $J_C = 0$, $K_C = 1$ and so the sixth clock pulse toggles flip-flop A and resets flip-flop C to make all the outputs, Q_A, Q_B and Q_C, equal to 0. Thus the arrangement of Fig. 5.22 is a synchronous *divide-by-6* counter.

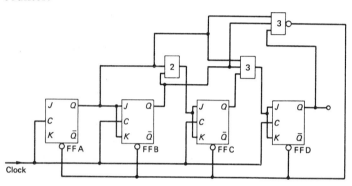

Fig. 5.23 Synchronous decade counter

An example of the reset method of count reduction is shown by Fig. 5.23. The 4-bit synchronous counter of Fig. 5.20 has been modified by the addition of a 3-input NAND gate. The output of this gate will be at 0 only when $Q_A = Q_B = Q_D = 1$, i.e. at the end of the *eleventh* clock pulse. Immediately the output of the gate goes to 0, the counter is reset and so the circuit counts through the sequence 0, 1, 2, 3, 4, 5, 6, 7, 8, 9, 10, 1, 2, 3, etc. Thus, Fig. 5.23 shows one form of synchronous decade counter.

Up-Down Counters

All of the counters described so far in this chapter have counted from 0 up towards some number and are hence *up-counters*. For some digital applications it is necessary to be able to count downwards, e.g. 9, 8, 7, 6, 5, 4, 3, 2, 1, 0. Many circuits are capable of counting in either direction and the basic arrangement of a non-synchronous *up-down* counter is shown in Fig. 5.24.

If the count-up line is taken to logical 1 level, the AND gates A and D are enabled, connecting the Q outputs of flip-flops A and B to the clock input of the following flip-flop. The circuit then operates as an up-counter having a count of 8.

When the count-down line is at logical 1, and the count-up line is at logical 0, gates B and E are enabled, while gates A

Count up

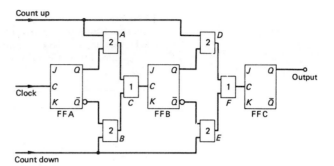

Clock

Output

FFA   FFB   FFC

Count down

Fig. 5.24 Up-down counter

and D are inhibited. Now the $\bar{Q}$ outputs of flip-flops A and B are connected to the clock inputs of the following stages. Suppose that initially all three flip-flops are set, i.e. the count is 7. At the end of the first clock pulse, flip-flop A resets so that $\bar{Q}_A = 1$ and the count is 1 1 0 or 6. The next clock pulse causes flip-flop A to toggle and the trailing edge of its $\bar{Q}$ pulse resets flip-flop B; now the count is 1 0 1 or 5. At the end of the third clock pulse, the first stage toggles so that $Q_A = 0$, $\bar{Q}_A = 1$ and the count is 0 1 0 or 4. The state of the counter is now $\bar{Q}_A = \bar{Q}_B = 1$, $\bar{Q}_C = 0$, and so the fourth clock pulse sets flip-flops A and B and resets flip-flop C to give a count of 3, and so on until all three stages are reset. The count is then 0 and the next clock pulse will return the counter to its original count of 7.

Most integrated circuit up-down counters are of the synchronous type, e.g. 74190/1/2/3 which are, respectively, b.c.d., binary, decade, and 4-bit types, but their internal circuitry is too complex to be outlined in this book.

C.M.O.S. Counters

The examples given so far of integrated circuit counters have all been devices in the t.t.l. family. Both ripple and synchronous counters are, however, also available in the c.m.o.s. logic family.

The relative merits of t.t.l. and c.m.o.s. circuits have been tabulated earlier (p. 45). T.T.L. counters can operate at much higher clock frequencies than c.m.o.s., e.g. standard t.t.l. 20–30 MHz, low-power Schottky t.t.l. 10–100 MHz, c.m.o.s. 10 MHz.

C.M.O.S. counters dissipate much less power. Another advantage of c.m.o.s. devices is that a much larger number of stages can be provided within a standard d.i.l. i.c. package. The 4020 has 14 stages and the 4045 has 21 stages, for example. Because of the limitations on the number of package pins, only a few stages (usually one) can have an output that is externally accessible. The 4045 is intended for use as a timing circuit.

Exercises

(Assume trailing-edge triggered flip-flops unless otherwise stated.)

5.1. Determine the operating sequence of the counter shown in Fig. 5.25. Draw the waveform diagram.

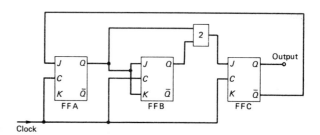

Fig. 5.25

5.2. Draw the diagram of a 3-bit synchronous counter that uses 3-input NOR gates for decoding the count.

5.3. Show how two 7493 t.t.l. binary counters can be connected to obtain a frequency division of 100.

5.4. (a) Explain what is meant by the term ripple counter. (b) Show how a ripple counter can be constructed using T flip-flops. (c) Show how a 4-bit binary counter can be made to give a count of 12. Assume each flip-flop has a clear terminal available.

5.5. (a) Draw the circuit of a divide-by-16 synchronous counter. (b) Draw the waveform diagram of your circuit assuming the flip-flops are triggered by the rising edge of each clock pulse.

5.6. Determine, with the aid of a truth table, the operation of the counter shown in Fig. 5.26. Draw its waveform diagram.

5.7. (a) Explain how successive stages of bistable multivibrators can be used to form a counter. (b) Explain the advantages to be gained by synchronizing the operation of the multivibrators. (c) What kind of bistable is most commonly used? (d) How should the inputs of this kind of bistable be connected so that it acts as a toggle?

5.8. Write notes on the various ways in which the count of a counter can be reduced below 2^n, where n is the number of stages in the counter.

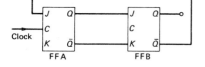

Fig. 5.26

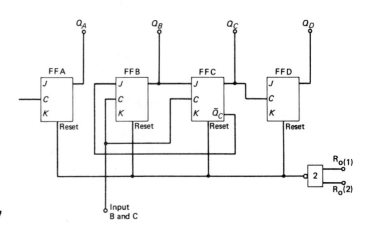

Fig. 5.27

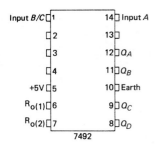

Input B/C ⊏1   14⊐ Input A

        ⊏2   13⊐

        ⊏3   12⊐ Q_A

        ⊏4   11⊐ Q_B

   +5V ⊏5   10⊐ Earth

$R_{o(1)}$⊏6   9⊐ Q_C

$R_{o(2)}$⊏7   8⊐ Q_D

         7492

(a)

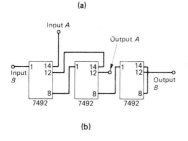

(b)

Fig. 5.28

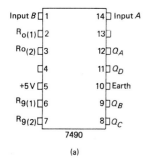

Input B ⊏1   14⊐ Input A

$R_{o(1)}$⊏2   13⊐

$R_{o(2)}$⊏3   12⊐ Q_A

       ⊏4   11⊐ Q_D

  +5V ⊏5   10⊐ Earth

$R_{9(1)}$⊏6   9⊐ Q_B

$R_{9(2)}$⊏7   8⊐ Q_C

         7490

(a)

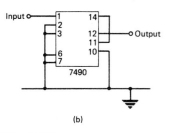

(b)

Fig. 5.29

5.9. Fig. 5.27 shows the internal connections of the t.t.l. 7492 counter. For the circuit to operate as a divide-by-12 circuit, Q_A is connected to the common B/C input terminal. Explain, with the aid of a truth table, the operation of the circuit when so connected.

5.10. (a) Explain, with the aid of waveforms and truth tables, the action of a J-K master-slave bistable element.

(b) Two J-K master-slave bistable elements are connected to form a counter. With the aid of a truth table, show how the combined elements can form a modulo-3 counter over 4 clock cycles.

5.11. Fig. 5.28a shows the pin connections of a t.t.l. 7492 counter; it consists of two separate sections providing, respectively, divide-by-2 from pin 14 to pin 12, and divide-by-6 from pin 1 to pin 8. Three 7492 counters are connected in the manner shown by Fig. 5.28b. Input A is at 12 kHz and input B is at 28 kHz. Determine the frequencies of the two outputs of the circuit.

5.12. The pin connections of a 7490 counter are given by Fig. 5.29a and the i.c. is connected as shown by Fig. 5.29b. Explain, with the aid of a truth table, the operation of the circuit.

Short Exercises

5.13. What is the maximum count obtainable from a ripple counter with (i) 3, (ii) 5 and (iii) 7 cascaded flip-flops?

5.14. Typical turn-on and turn-off times for a t.t.l. flip-flop are 75 ns. Calculate the propagation delay of a t.t.l. counter using (i) 3, (ii) 4 stages of flip-flops.

5.15. A ripple counter contains 4 flip-flops, A, B, C and D. How much faster does (i) A operate than C, (ii) B operate than D, (iii) B operate than C, (iv) C operate than D?

5.16. Fig. 5.12 shows the pin connections of a 7493 4-bit binary counter. Which pins should be connected together to produce a divide-by-5 circuit?

5.17. A decade counter is to be used in conjunction with a J-K flip-flop to provide a divide-by-20 circuit. How should the two devices be connected?

5.18. Why cannot the feedback method of reducing the count of a counter be applied to an integrated circuit version?

5.19. A 3-stage down-counter has all its stages reset. Explain how the next clock pulse will set all stages to give a count of 7.

5.20. How must a J-K flip-flop be connected in order for it to act as a toggle? Write down the truth table of a toggle.

5.21. Write down the truth table of a divide-by-16 counter. How many counts must be omitted if the counter is to be modified to divide-by-10?

6 Memories

Types of Memory

Many digital systems include a **memory** or **store** for the temporary or long-term storage of information. In a digital computer, this information will include numerical data, the intermediate results of computations, and the programmes which control the operation of the system. In a telephone exchange, a memory may be used to store code translations and information about each line, such as its number and the nature of the equipment connected to it. A memory should have sufficient capacity to be able to satisfy all the demands made for data and programme storage and should be so fast to operate that undue delay is not caused to the main processor. It is also desirable for a memory to be of the minimum possible cost and to be reliable and also for the stored data to be retained if the power supplies should be shut down. Because of the various demands on a memory a computer employs more than one kind of store.

A **main store** contains all the information which must be immediately available, such as the interim and final results of calculations. A larger-capacity **backing store** contains most of the data and the programmes and it is used to hold information that need not be instantly accessed by the central processor. Short-term memory is provided by **registers**. Registers are used to store the interim results of arithmetic processes and for the movement of data between the processor and the main memory. The *memory-address register* contains the address of the data held in the memory which is to be involved in a read, or a write operation. Registers are also involved in the movement of data between the processor and the input/output devices.

It must be possible both to write information into a memory and to read information out of the memory. To make this possible a memory consists of a large number of **locations** in each of which a small amount of data can be stored. Each location has a unique **address** so that it can be accessed from outside the memory. The **access time** of a memory is the time that is needed to read one *word* out of the memory, or to write one word into the memory. The access time of a main store must be measured in nanoseconds and this means that it must be of the random access type. A **random access memory**, or ram, is one in which any location can be accessed without having to go through all the addresses in numerical order.

Thus the time taken to read from, or to write into, any location is the same as for any other location. Rams can be manufactured using large numbers of ferrite cores and are also available in integrated circuit form.

The backing store is required to hold a large quantity of data at the minimum cost and it is generally either a magnetic tape or a magnetic disc. Data can only be read out of or written into a tape or a disc as the appropriate part of the tape (or disc) moves beneath the read and/or write head and this means that the access time can be relatively long.

The basic requirements of any memory or store are that

(*a*) Any required location in the store can be addressed.
(*b*) Data can be read out of an addressed location.
(*c*) Data can be written into an addressed location (once only in some cases).

If a memory is able to retain the data stored in it when the power supplies have been switched off, the memory is said to be **nonvolatile**. A volatile store will lose the data stored within it if the power supplies are removed. Magnetic memories are nonvolatile and semiconductor memories are volatile.

Read-only memories, or roms, are memories into which data is permanently programmed either at the time of manufacture or by the user prior to the memory being installed into an equipment. Essentially a rom consists of a matrix of conductors some of whose intersections are joined by diodes or transistors.

The Shift Register

A shift register consists of a number of J-K flip-flops connected in cascade as shown by Fig. 6.1. The first stage is connected as a D flip-flop. The number of flip-flops used to form the register is equal to the number of digits to be stored. The flip-flops are generally provided with clear or reset terminals so that the register can be cleared.

Suppose that initially all the four flip-flops shown are

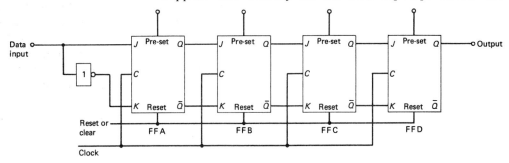

Fig. 6.1 Shift register

Fig. 6.2 Movement of data through a shift register

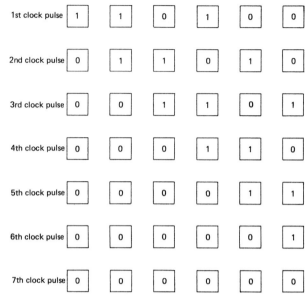

1st clock pulse	1	1	0	1	0	0
2nd clock pulse	0	1	1	0	1	0
3rd clock pulse	0	0	1	1	0	1
4th clock pulse	0	0	0	1	1	0
5th clock pulse	0	0	0	0	1	1
6th clock pulse	0	0	0	0	0	1
7th clock pulse	0	0	0	0	0	0

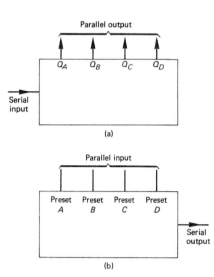

(a)

(b)

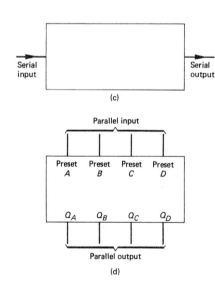

(c)

(d)

Fig. 6.3 Methods of using a shift register:
(a) serial-in/parallel-out,
(b) parallel-in/serial-out
(c) serial-in/serial-out
(d) parallel-in/parallel-out

cleared, i.e. $Q_A = Q_B = Q_C = Q_D = 0$ and that the 4-bit word to be stored is 1011.

At the end of the first clock pulse, flip-flop A is set so that $Q_A = J_B = 1$.

After the second clock pulse flip-flop B is also set, and now $Q_A = Q_B = 1$.

The state of the register when the third clock pulse arrives is $J_A = 0, K_A = 1, J_B = J_C = 1, K_B = K_C = 0$ and at the trailing edge of the clock pulse flip-flop A resets and flip-flop C sets. Now $Q_A = J_B = 0,\ Q_B = Q_C = J_C = J_D = 1$.

The last bit to be stored is 1 and at the end of the fourth clock pulse flip-flops A, C and D are set and flip-flop B is reset. If no more data is applied to the input terminals of the register, four more clock pulses will return the register to its original reset condition.

The effect of each clock pulse is to shift the content of each stage one place to the right, with flip-flop A storing the data at the input terminals of the register. This action is illustrated by Fig. 6.2 in which it has been supposed that the data 1101 has been fed into a 6-stage shift register. Some shift registers are arranged to operate in the opposite direction, that is the stored data shifts one place to the left each time a clock pulse is applied.

A shift register can be used in any of four ways:
   (i) Serial-in/Parallel-out
   (ii) Parallel-in/Serial-out
   (iii) Serial-in/Serial-out
   (iv) Parallel-in/Parallel-out

With a **serial-in/parallel-out** register (Fig. 6.3*a*) data is fed into the circuit in the manner just described and, when the complete word is stored, all the bits are read off simultaneously from the output of each stage. The register acts to convert data from serial into parallel form.

The **parallel-in/serial-out** register (Fig. 6.3*b*) operates in exactly the opposite way. The data to be stored is set up by first clearing all the stages and then applying a 1 to the preset terminal of each flip-flop which is to be set. The data is read out of the register, one bit at a time, under the control of the clock.

The **serial-in/serial-out** register (Fig. 6.3*c*) can be used as a delay circuit or as a simple store but the stored data can only be accessed in the order in which it was stored.

Fig. 6.3*d* shows a **parallel-in/parallel-out** register and this also acts as a short-term store.

Sometimes there is a need for a shift register that has the capability to move data *either* to the left *or* to the right. The circuit of such a register is shown in Fig. 6.4. There are two data input terminals, one of which is used for serial data that is to be shifted to the right, and the other is for left-shifting data. The direction in which the data is shifted is determined by the logic levels on the two lines marked, respectively, as "shift-right-1/ shift-left-0" and "shift-right-0/shift-left-1". The circuit is arranged so that the signals applied to these two lines are always the complements of one another. When the top AND gates are enabled the data right input is connected to flip-flop A, Q_A is connected to flip-flop B, and so on, so that the circuit is similar to that given in Fig. 6.1 (p. 77).

Conversely, when the lower AND gates are enabled, Q_D is connected to the D terminal of flip-flop C, Q_C is connected to flip-flop B, and so on. The circuit will then shift data entered serially at the data left terminal to the left.

Fig. 6.4 Shift-left/shift-right register

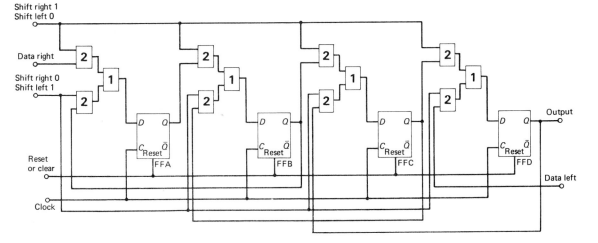

A shift register can be constructed using a number of integrated flip-flops but more commonly a t.t.l. or a c.m.o.s. shift register would be used. C.m.o.s. shift registers are generally only of the serial-in/serial-out type mainly because they contain many stages (e.g. the 4006 is an 18-stage register) and there are not enough package pins available for parallel input and/or output to be possible. The capacity of an integrated shift register ranges from 4 bits in the t.t.l. family to 2048 bits in the c.m.o.s. family.

T.T.L. shift registers hold the same speed advantage over their c.m.o.s. equivalents as discussed for counter circuits. The capacity of a t.t.l. register is limited to 8-bit words but all four types of input and output are available. Some c.m.o.s. shift registers are only four-stage or eight-stage circuits, and their advantage therefore lies with their low power dissipation. These registers are available with all four input/output combinations. Also available in the c.m.o.s. family are shift registers with 18, 32, 64, and even 200 stages. Because of i.c. package pin limitations, such circuits are serial-in/serial-out devices only. In most families most registers are right shift only but some circuits provide both left and right shift facilities.

The Ferrite Core Store

Magnetic stores are used as the main store in digital computers and other digital systems, and very often they use a number of *ferrite cores.* The ferrite core store is popular because it provides fast access but it does occupy rather a lot of space, it is not cheap, and when a large capacity store is required it is necessary to provide backing storage as well. The backing storage is generally magnetic tape or magnetic disc although in older equipments the magnetic drum is also employed. These backing stores are all of much greater storage capacity than the ferrite core store but they share the common disadvantage of slow access time. In this book attention will be paid to the ferrite core store only.

Ferrite is a manufactured magnetic material that has a high permeability, a high resistivity, and a square hysteresis loop as shown by Fig. 6.5. When a magnetizing force H is applied to the material, the magnetic flux density is increased to the saturation level of B_{sat}. If the magnetizing force is then removed, the flux density does not fall to zero but instead falls only very slightly to a value, known as the *remanent flux density* B_r, which is almost as large as the saturation flux density. Further, the flux density will hardly change from this value even if the direction of the magnetizing force is reversed, unless the magnitude of this reversed m.m.f. is equal to the critical value, labelled as H_s in Fig. 6.5. Immediately the

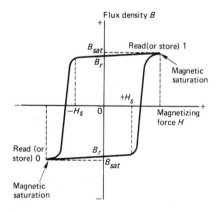

Fig. 6.5 Hysteresis loop of a ferrite core

reversed m.m.f. reaches its critical value the flux density in the ferrite core rapidly changes state to its saturation value in the opposite direction.

This means that, once a ferrite core has been magnetized in one direction or the other, it will remain in that state until such time as a sufficiently large magnetizing force is applied to reverse the direction of magnetization. A ferrite core is therefore an example of a two-state device and it is therefore able to store either of the two binary digits at a time. Usually the positive saturation state is used to represent binary 1 and the negative saturation state represents binary 0. The hysteresis loop must be as near rectangular in shape as possible to ensure that any small changes in m.m.f. do not result in much, if any, change in the flux density. Otherwise the flux density may gradually reduce until the core is no longer saturated.

For the logical state of a ferrite core to be changed, a magnetizing force of magnitude H_s must be applied in the opposite direction. Since magnetizing force is equal to current times number of turns, this means that a current of suitable magnitude must be passed through a winding on the core. This is shown by Fig. 6.6 from which it can be seen that the "winding" consists of a single wire that passes through a hole in the centre of the core. A single wire is used because of the very small dimensions of a ferrite core, typical dimensions being outside diameter 0.5 mm, diameter of hole 0.3 mm.

To *read* the logical state of a core two further "windings" are needed as shown by Fig. 6.7. A pulse of current is passed through each of the two **read wires** in directions such that the total magnetic flux created will reset the core to the saturation condition that represents binary 0. If the core was already in the logical 0 state the core flux will not change. If a change in the direction of core magnetization does occur, the change in flux will induce an e.m.f. into the **sense wire**. Thus, when a core is read, a voltage pulse will appear on the sense wire only if the stored bit was 1; if the stored bit was 0, zero voltage will appear on the sense wire.

The readout process always resets the core to the logical 0 state, any 1 bit being lost. The readout is said to be **destructive**. Usually it is required to keep the stored data and so provision must be made for the read-out data to be re-written after readout has taken place.

Structure and Operation

A ferrite core store consists of a **stack of matrices**, or **planes**. In each plane of a stack, a single bit is stored and so the number of planes must be equal to the number of bits in the word to be stored. A *word* is a number of bits which are

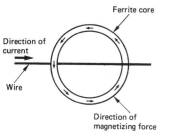
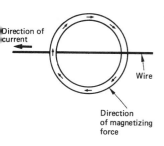

Fig. 6.6 Magnetizing a ferrite core

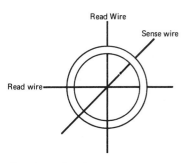

Fig. 6.7 Ferrite core with read and sense wires

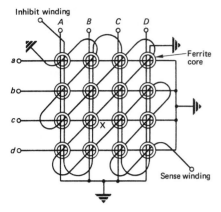

Fig. 6.8 16-core matrix

grouped together to form a unit that can be given a unique location in the store.

Fig. 6.8 shows how 16 ferrite cores are arranged in a 4×4 matrix. The 16 cores are positioned in four rows of four columns and each of the 16 cores has four wires passing through it. Each core in the same row is threaded by a common row wire, labelled a, b, c or d, and each core in the same column shares a column wire, labelled A, B, C or D. Also, a sense wire is threaded through all 16 cores following a zig-zag path as shown. This particular wiring arrangement is used because it allows any core in the matrix to be read from or written into.

Suppose that initially all the 16 cores are in the logical 0 state. For any one of the cores to be switched to store a 1 bit, pulses of current must be applied to both the associated row and column wires. These current pulses must each have an amplitude equal to one-half of the current needed to produce the critical value of magnetizing force that will initiate switching. If, for example, the core marked X in Fig. 6.8 is to be switched, current pulses should be simultaneously applied to row wire c and to column wire B only. The other cores in row c and in column B only have one-half the critical magnetizing force applied to them and so they do not change state (see Fig. 6.9). Fig. 6.9a shows the hysteresis loop of core X; the magnetizing forces produced by the column and row currents add to give the critical value H_s of magnetizing force needed to switch a core. The other cores in row c and in column B have the hysteresis loop shown in Fig. 6.9b; the magnetizing force developed is due to *either* the c row current *or* the B column current and is not large enough to cause the cores to

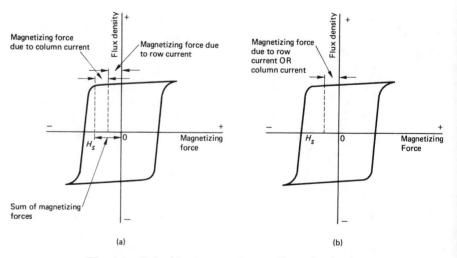

Fig. 6.9 Coincident current operation of a ferrite core

switch. Since only one core, the one whose row and column wires *both* carry current, is set, this method of operation is known as **coincident current operation.**

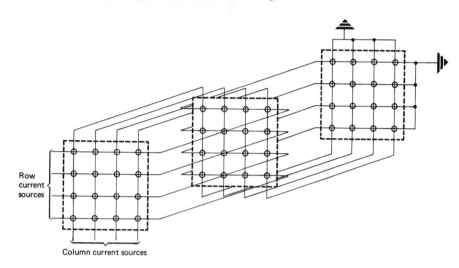

Fig. 6.10   A 3-plane 4×4 matrix ferrite core store

The fourth wire, known as the **inhibit wire**, is used when data is written into the store. The inhibit wire passes through each core in parallel with a column wire. If, therefore, the inhibit wire carries a current, of equal magnitude but in the opposite direction, to the current in the column wire, the magnetic fields set up by these two currents will cancel. Fig. 6.10 shows a simple store consisting of a stack of three 4×4 matrices or planes. To write a word into the store, **half-write currents** are simultaneously passed along one row and one column wire to select one core in each plane. This will set each of the selected cores to the logical 1 state *unless* a current is also passed through the inhibit wire of a plane; then the core in that plane will remain in the 0 state. If, for example, the word to be stored is 110, an inhibit current will be caused to flow in the third plane only.

The capacity of a store is normally quoted in terms of **kilobits**. The prefix *kilo* does not stand for 1000, as is usual, but for 2^{10} or 1024. Thus

60 kilobits = 60 K = 60 × 1024 = 61 340 bits

EXAMPLE 6.1

How many ferrite cores are there in a 12 kilo, 16 bit word length store?

Solution   Number of cores = 12 × 1024 × 16 = 196 608

Semiconductor Memories

Semiconductor memories have now become the predominant memory technology because they are cheaper, smaller, and faster operating than the magnetic alternatives. A complete random access memory (ram) or read only memory (rom) can be formed within a single integrated circuit chip and made available in a standard d.i.l. package.

A memory consists of a matrix of memory cells together with digital circuits that provide such functions as *address* selection and *control*. Each memory cell is situated at a matrix location that is identified by a unique address.

Random Access Memories

The matrix of a **random access memory** or **ram** is organized as m words of n bits each, i.e. $m \times n$. The memory cells are located at the intersections of the m rows and the n columns of the matrix. The idea is illustrated by Fig. 6.11a which shows a square matrix in which $m = n = 5$. This is not a size used in practice but has been drawn for simplicity. Practical rams may use a square matrix, for example 32 words of 32 bits (1024 bits or 1 kilobit) but often employ rectangular matrices; typical examples are 64×16 or 256×4 (both 1 kilobit).

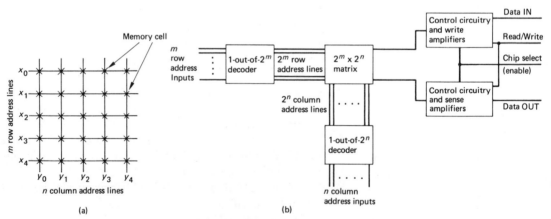

Fig. 6.11 Random access memory: (*a*) memory matrix, (*b*) block diagram

To reduce the number of column and row address lines needed, addresses are signalled using the binary code and are decoded to produce voltages on the selected column and row lines. The arrangement used is shown in Fig. 6.11b. Suppose, for example, that the memory cell located at the intersection of row 2 and column 3 is to be selected. Then the input to the row decoder must be 010 and the input to the column decoder must be 011.

The diagram also shows blocks marked as control circuitry/ write amplifiers and control circuitry/sense amplifiers. These circuits perform the functions either of writing new data into a location, or of reading out existing data. The read/$\overline{\text{write}}$ input determines which of the two functions is performed. The chip select or enable input must (usually) be high to enable the memory; this facility is used when two or more chips are combined to produce a larger capacity memory.

When the state, 1 or 0, of a memory cell is to be read, the read/$\overline{\text{write}}$ line is set to the logical 1 voltage level, and the address of the required location is fed into the address decoders. The data held at that location then appears at the data-out terminal(s). The read-out process is *non-destructive*.

When new data is to be written into the memory, the read/$\overline{\text{write}}$ line is set to the logical 0 voltage level. The data present at the data-in terminal(s) will then be written into the addressed memory cell(s).

Two kinds of ram are manufactured, known respectively as the static ram and the dynamic ram. In a **static ram** the memory cells are actually flip-flops that can be fabricated using either bipolar transistor or m.o.s.f.e.t. technology. Examples of static rams in the t.t.l. family are the 74170 4×4 and the 74670 4×4, the former having a totem-pole output and the latter having a three-state output. M.O.S. static rams are of larger capacity, such as 4096×1 or $16\,384 \times 1$, and dissipate less power than the bipolar transistor versions but, on the other hand, they have greater access times.

A **dynamic ram** does not use flip-flops as the memory cells but, instead, data is stored in the stray capacitances that inevitably exist between the gate and the source of a m.o.s.f.e.t. The dynamic ram has the advantages that (i) a larger storage capacity can be provided within a given chip area, (ii) it is faster to operate and is cheaper to fabricate than a static ram. On the other hand the dynamic ram possesses the disadvantage of needing periodic *refreshing* of the stored data.

A large number of static rams are available in the t.t.l. family; among the memory capacities provided are 16×4, 16×9, 16×12, 32×8, 64×4, and 256×1.

Dynamic rams can only be provided using m.o.s.f.e.t. technology, examples are $4K \times 1$ and $16K \times 1$.

Read Only Memories

A **read only memory** or ɹom has data written into it, in *permanent form*, by either the manufacturer or the user. When in use, data can *only* be read out of the memory; new data *cannot* be written in. A rom is non-volatile.

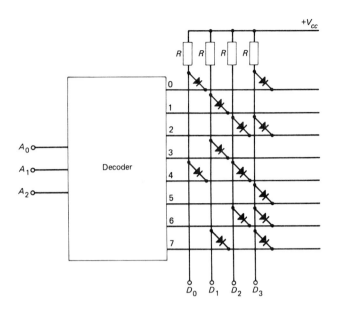

Fig. 6.12 Diode read only memory

The organization of a rom is very similar to that of a ram. Data is stored at different locations within the memory matrix and each location has a unique address. When a particular location is addressed, the data stored at that address is read out of the memory. The read-out is non-destructive.

The arrangement of a diode rom is shown in Fig. 6.12. Diodes are connected between some of the row and some of the column lines. When a decoder line, 0 through to 7, goes low, the associated diodes turn ON. Each output line D_0, D_1, D_2 or D_3 connected to an ON diode is then taken low also. When for example, the input address is $A_0 = 0$, $A_1 = A_2 = 1$, line 6 goes low and the output of the rom is $D_0 = D_1 = 1$, $D_2 = D_3 = 0$. The Boolean equation describing this action is

$$F = \bar{D}_0 D_1 D_2 \bar{D}_3 + D_0 \bar{D}_1 D_2 D_3 + D_0 D_1 \bar{D}_2 \bar{D}_3 + \bar{D}_0 D_1 \bar{D}_2 D_3 \\ + D_0 D_1 D_2 \bar{D}_3 + D_0 D_1 \bar{D}_2 \bar{D}_3 + D_0 \bar{D}_1 D_2 \bar{D}_3 + D_0 \bar{D}_1 D_2 D_3$$

Integrated circuit roms may use either bipolar or field effect transistors as the row and column linking elements.

Typical applications for a rom are code conversion, mathematical tables and programme controllers.

A rom is programmed during manufacture and this data cannot be subsequently altered. This means that the intending user must inform the manufacturer of the particular data that each location is to contain. This is alright for the large-scale user but it is much less convenient for the user of much smaller quantities. To provide some flexibility in the possible applications of roms, programmable devices are generally used.

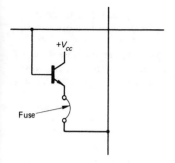

Fig. 6.13   Element in a prom

Programmable ROMs

A **programmable read only memory** or **prom** is designed so that it can be programmed by the user to suit a specific application for the device. All of the intersections in the memory matrix are linked by a fusible diode or transistor (see Fig. 6.13). When the prom is purchased from the manufacturer, all of the outputs are at the logical 0 voltage level. The programming procedure consists of changing the bits stored at selected locations from 0 to 1.

Programming of a prom is accomplished by addressing a particular location in the memory that is to store a 1, and then passing a sufficiently large current through the transistor to blow the fuse. The transistor then no longer links the row and column lines at that location.

There are several t.t.l. proms available, some examples being 74186 64×8, 74199 32×8, 74287 254×4, and 74470 256×8. E.C.L. devices are also available.

Proms are widely used in the control of electrical equipment such as washing machines and ovens.

Erasable and Electrically Alterable PROMs

Some proms can have their programmes altered and a new programme written into the memory. In an **erasable prom** or **eprom** the logical 1 state is stored at a location by the storage of an electrical charge and not by the blowing of a fuse. When a programme is to be erased, the chip is exposed to ultra-violet radiation that is directed through a window in the chip package. This radiation removes the stored charge at *every* location in the memory so that all locations store binary 0. Re-programming is carried out by addressing each cell that is to store a logical 1 bit and then causing that cell to store a charge.

An alternative to the eprom is known as the **electrically alterable prom** or **eaprom**. Again, programming a memory cell to store logical 1 is accomplished by charging that cell. With the eaprom, however, the erasure procedure is carried out by applying a reverse-polarity voltage to a cell that removes any stored charge. The eaprom offers an advantage over the eprom in that the erasure process can be applied to an individual cell in the matrix if required.

Both eproms and eaproms are m.o.s.f.e.t. devices but are not in the c.m.o.s. family.

Exercises

   6.1. Fig. 6.14 shows the pin connections of a 7476 dual J-K flip-flop. Draw a diagram to show how two such integrated circuits could be connected to form a 4-stage shift register. Write down the truth table of the register.

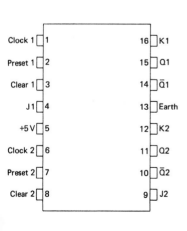

Clock 1	1	16	K1
Preset 1	2	15	Q1
Clear 1	3	14	$\bar{Q}1$
J1	4	13	Earth
+5 V	5	12	K2
Clock 2	6	11	Q2
Preset 2	7	10	$\bar{Q}2$
Clear 2	8	9	J2

Fig. 6.14

6.2. (a) Draw the hysteresis loop for a typical ferrite core and describe how this magnetic effect can be used to store a binary digit.

(b) Why must the loop be as rectangular as possible?(C & G)

6.3. A magnetic core store contains 256 locations of 6-bit words. With the aid of a suitable diagram explain the operation of such a store when it is required to read the binary number 101010 from a selected location and re-write that same value into the same location. (C & G)

6.4. With the aid of block and timing diagrams explain the mode of operation of (a) a 3-bit counter, (b) a 3-bit shift register.

6.5. Define the terms static ram and dynamic ram. List the advantages and disadvantages of a bipolar transistor static ram compared with a m.o.s.f.e.t. static ram.

6.6. What is meant by the terms volatile and non-volatile when applied to a memory or store? Which type is (i) a ram, (ii) a rom? Give one example of the use of each type of memory.

6.7. The c.m.o.s. 4731 is a 64-bit shift register. (a) How long a word can·this register store? (b) Will the circuit be operated in a parallel or serial input and/or output mode? (c) How long a time delay will this circuit introduce if the clock frequency is 10 kHz? Give reasons for each of your answers.

6.8. With the aid of logic diagrams explain how three clocked flip-flops can be used to form (i) a binary counter, (ii) a shift register.

Short Exercises

6.9. Explain the difference between a ram and a rom. List the input and output signals of each memory.

6.10. A 64-bit semiconductor square matrix is addressed by the binary number 101011. In which row and which column is the wanted flip-flop located?

6.11. Why must the hysteresis loop of a ferrite core be as near rectangular as possible?

6.12. Why is a semiconductor memory said to have a non-destructive readout?

6.13. Describe briefly how a ring counter can be made from a shift register.

6.14. Sketch the hysteresis loop of a ferrite core and label the important operational points.

6.15. Explain how a flip-flop may be used to store a binary digit.

6.16. State briefly the purpose of the read, write, and sense wires in a ferrite core.

6.17. What is meant by the term coincident current switching when applied to a ferrite core?

6.18. What are the advantages and disadvantages of ferrite cores as rams compared with the semiconductor alternatives?

7 Data Communication and the Transmission of Pulses over Lines

Types of System

Many firms and organizations have a number of factories, warehouses, offices and other points where members of staff are employed. Usually, many of these locations are not sited at the headquarters, some indeed may be sited many hundreds or even thousands of kilometres distant. All of the remote locations need to send information to the headquarters at regular intervals varying from weekly to daily, or even shorter. The information can, of course, be sent by post but there is an ever increasing demand nowadays by management for information to be made available more quickly. The answer to the problem can very often be given by telegraphic techniques, in particular the Telex system (TS 1), but this is limited in its maximum speed of transmission, and some faster system is often required. Much of the data required by a head office is destined to be stored and/or processed by a digital computer, and computers are employed for a great number of varied purposes in both the engineering/scientific and the commercial fields. Computers are used to carry out complex calculations, to maintain and update records such as technical data, bank accounts and medical histories. Computers are also used in the preparation of weather reports and forecasts, the calculation and printing of wages and salaries, and of bills and invoices; they are also used for airline and package holiday bookings and reservations, for inventory control, and for the control of industrial processes in factories.

The many applications of the digital computer have led to most, if not all, of the larger firms and organizations investing in their own computer facilities. However, a digital computer is an expensive purchase and it is not economically possible for an organization to install computers at several locations. Yet the facilities provided by the computer may be needed at many different points in the organization's set-up. Because of this there is a considerable demand for communication links, or **data links**, to connect **data terminals** with the main **computer centre** and so extend the use of the computer installed at the centre to a number of distant locations.

Data can be collected at a branch office, transmitted to the computer centre, processed by the computer, and then either stored or transmitted back to the branch office. The data system can be operated as either an on-line or an off-line system. In an **on-line system** the data transmitted from a branch office appears to be fed directly into the computer. With an **off-line system** the branch office data terminal transmits its data to another terminal within the computer centre (possibly by hand or by post), then the data is stored on a card or a tape, and is fed into the computer at some later time.

An on-line system may also be said to operate in **real time**. This term means that the computer acts on the input data immediately and the computer output is provided within a very short time. An example of a real time system occurs with the package holiday firms; when a booking clerk makes contact with the computer and asks it about the availability of a particular holiday, an answer is needed within a few seconds. If the holiday enquired about is still available and is to be reserved for the customer then and there, the computer's stored information must be updated to reserve that holiday and take it off the availability list to prevent double booking. The use of a real-time computerized booking system makes the immediate confirmation of holidays possible.

Much of the data transmitted over data links originates from a keyboard printer of some kind or another, or from a paper tape or card reader, but a variety of another data terminal equipments are also used including **visual display units** (v.d.u.s). Several examples of data systems are given in Chapter 9 but the general principle of a system is shown by Fig. 7.1.

Communication from a point within or close to the computer centre to the computer can be carried out using a keyboard printer or a paper tape reader/puncher which is **directly linked** to the computer by a short ±6 V signal link. Longer length data links can also use d.c. signals but the transmitted voltage must be increased to ±80 V in order to offset the effects of line attenuation. Such signals cannot pass through telephone exchange switching equipment for reasons mentioned later and therefore a long ±80 V circuit must be leased permanently from the telephone administration. Unless low-speed working is satisfactory, the maximum distance possible is limited to a few kilometres. A **public switched telegraph network**, known as telex in the United Kingdom, is also generally available but the maximum speed of transmission is limited to 110 bits/second. In most cases data links to the more distant locations employ an equipment known as a **modem** to convert the ±6 V signals produced by the computer input/output devices into a voice-frequency signal that can readily be transmitted over the telephone network. A variety of modulation

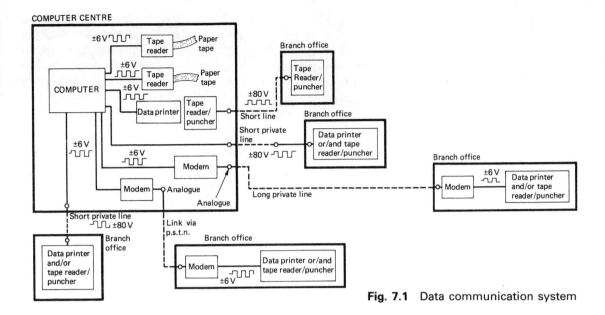

Fig. 7.1 Data communication system

methods are used in modems and these will be discussed in Chapter 8.

The longer distance data links can be set up by dialling a connection via the **public switched telephone network** (p.s.t.n.), or by leasing a permanent private circuit from the telephone administration. A leased private circuit is often called a *dedicated circuit*. A digital computer works at a very much faster speed than its input/output devices are able to, and so several users can be connected to the computer at the same time and yet each have the impression that he has exclusive use of the computer. Because of this, a communication controller and/or multiplexor is generally fitted in front of the computer to control the connection of different users/data links to the computer.

Single-current and Double-current Working

The data signal that is transmitted to line can be either a *single-current* or a *double-current* signal. **Single-current operation** means that only one polarity of voltage is used, and data is signalled by the application of, and the removal of, the voltage applied to the sending end terminals of the line. When **double current working** is used, two voltages are employed, one being positive with respect to earth and the other negative. Both methods of signalling can be operated over a single-conductor circuit using earth return or over a pair of conductors.

A data circuit can be operated in the *simplex*, the *half-duplex*, or the *full-duplex* modes. **Simplex** means that transmission of data is possible in only one direction over the link. A **half-duplex** circuit can be operated to transmit in either direction *but* only in one direction at a time. Lastly, a **full-duplex** circuit is capable of simultaneously transmitting data in both directions.

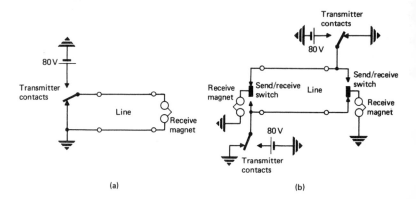

Fig. 7.2 (*a*) Single-current simplex circuit, (*b*) Single-current half-duplex circuit

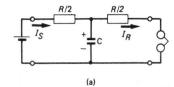

(a)

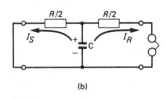

(b)

Fig. 7.3 Low-frequency equivalent circuit of a transmission line, showing the currents flowing for single-current operation

The basic principle of a **single-current simplex** circuit is shown by Fig. 7.2*a*. When the keyboard of the printer is touched, the transmitter switch operates and connects the +80 V battery to the line. Current then flows from the battery into the line and then through the receiving apparatus, before returning via the other conductor to the battery. The data is signalled by means of an interrupted d.c. voltage applied to the line. At zero and low frequencies the inductance and the leakance of the line both have negligible effect upon the signal and only the line's resistance and capacitance will affect the waveform of the received current.

Fig. 7.3 is an approximate representation of a low-frequency line which assumes that the total shunt capacitance is concentrated at the centre of the line. When the +80 V battery is connected to the sending end of the line, most of the current which flows into the line is used to charge the line capacitance. The current arriving at the far end of the line increases at a relatively slow rate and it is unable to reach its final *steady-state* value until the line capacitance has been fully charged. Once the line capacitance is fully charged, the sending-end current falls until it reaches the value needed to maintain the received current at its steady-state value. The time required for the receive current to reach its steady-state value is pro-

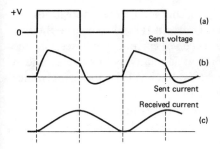

Fig. 7.4 Single-current waveforms

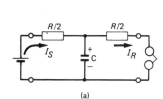

(a)

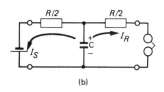

(b)

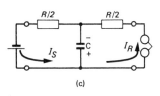

(c)

Fig. 7.6 Low-frequency equivalent circuit of a transmission line showing·the currents flowing for double-current operation

portional to the *total* capacitance C and resistance R of the line. Similarly, when the battery voltage is removed from the sending-end of the line and the terminal is earthed, the line capacitance will start to discharge at a rate determined by the time constant CRl^2, where l is the length of the line in kilometres. The discharge current is in the *opposite* direction to the original sending-end current, hence the sending-end current reverses its direction (see Fig. 7.3b). Some of the discharge current flows out of the receive end of the line in the *same* direction as the earlier current and so prolongs the receive current. Because of this the receive current does not fall as rapidly as the sent current, neither does it reverse its direction. Fig. 7.4 gives the waveforms of (a) the sent voltage, (b) the sent current, (c) the received current.

Fig. 7.2b shows the arrangement of a **single-current half-duplex** circuit; the operation is very similar to that just described except that transmission is possible in either direction.

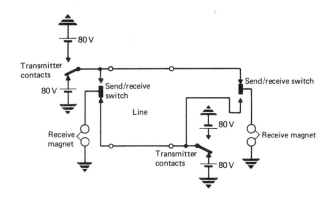

Fig. 7.5 Double-current half-duplex circuit

Fig. 7.5 shows the basic circuit of a **half-duplex double-current circuit**. The operation of the transmitter contacts reverses the polarity of the voltage supplied to the sending end of the line. Suppose a positive voltage, with respect to earth, is first applied to the line (Fig. 7.6a). When the line capacitance has been fully charged, the received current will reach its steady-state value after a time determined, as before, by the time constant CRl^2.

When the polarity of the sending-end voltage is first reversed (Fig. 7.6b), the line capacitance will start to discharge through both the transmitter and the receiver. The polarity of the reversed applied voltage is such that it acts in the same direction as the voltage developed across the line capacitance. As a result the capacitance of the line discharges much more rapidly than in the single-current case. This means that the receive current falls more rapidly. Once the line capacitance

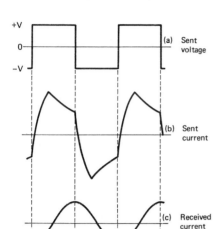

Fig. 7.7 Double-current waveforms

has been discharged, it is then charged in the opposite direction; most of the sending-end current is used to charge the line capacitance and thus the receive current increases, in the opposite direction to before, at a relatively slow rate, and it does not reach its steady-state value until the line capacitance has been fully charged (Fig. 7.6c).

Waveforms of current and voltage for a double-current circuit are given in Fig. 7.7. Clearly, the build-up and the decay times of the receive current waveform are smaller than in the single-current case. Further, the amplitude of the received current is greater.

Relative Merits of Single-current and Double-current Working

Single-current operation of a d.c. data circuit requires only one polarity voltage supply, whereas double-current operation requires the provision of both positive and negative power supplies.

If the *bit rate* (see page 91) of the data waveform is too high, the time taken for the receive current to reach its steady-state value may exceed the duration of the pulse. Then excessive pulse distortion will occur, making it difficult for signals to be reliably received. Double-current operation of a data link increases the rate at which the received current rises towards its final value and so allows a higher bit rate to be employed. Double-current working also results in a larger-amplitude received current than does the single-current system for the same battery voltage. This gives more reliable operation.

A further disadvantage of the single-current method of operation is that any momentary break in the transmission path will not be detected as such but will be recorded as binary 1 to produce an error in the received data.

The Arrival Curve

The **arrival curve** is a graph of the received current at the end of a line plotted against time for each positive or negative pulse taken separately, ignoring for simplicity any propagation delays (as was done in the previous figures). The arrival curve for a single pulse is shown by Fig. 7.8. The arrival curve can be used to construct the waveform of the received current for any input data waveform.

Each input voltage pulse is assumed to produce a current at the receiving end of the line corresponding to the arrival curve. The arrival curve for each input pulse is separately drawn and the waveform of the actual received current is then deduced by adding algebraically the individual arrival curves.

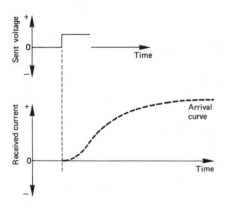

Fig. 7.8 Arrival curve for a single pulse

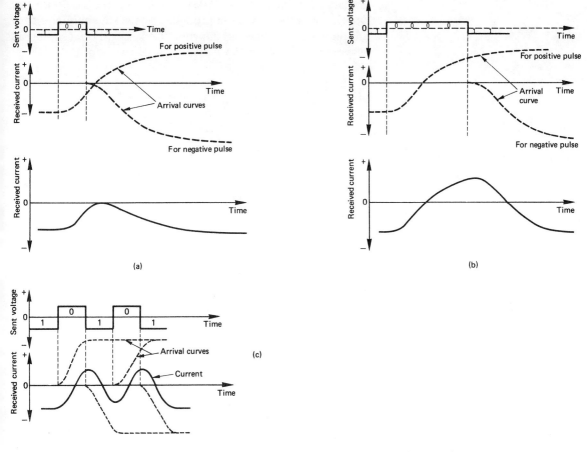

Fig. 7.9 Examples of the use of the arrival curve

The sketches given in Fig. 7.9 apply this principle to, first, two pulses of differing time durations and, secondly, a square waveform, both signals being centred on 0 V.

The Frequencies Contained in a Data Waveform

The data fed into and out of a computer use ±6 V pulses to represent the binary numbers 0 and 1 as shown by Fig. 7.10. Each signal element or *bit* has the same time duration in seconds as all the other bits and the number of bits transmitted per second is known as the **bit rate**, or the **data signalling rate**. If, for example, the bit duration in Fig. 7.10 is 9.09 ms, the bit rate would be $1/9.09 \times 10^{-3}$ or 110 bits/sec.

Fig. 7.11 shows a sinusoidal voltage $v = V \sin \omega t$, of peak value V and frequency $\omega/2\pi$, and another, smaller, voltage at three times the frequency, i.e. $3\omega/2\pi$. The first voltage is the fundamental frequency and the other voltage is its third har-

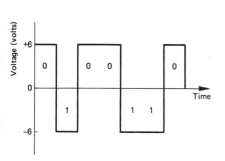

Fig. 7.10 Data waveform

monic. At time $t = 0$, the two voltages are in phase with one another and the waveform produced by the summation of their instantaneous values is shown dotted; clearly the resultant waveform is tending towards a rectangular shape.

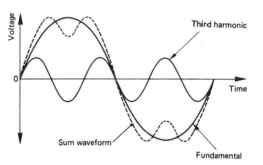

Fig. 7.11 Waveform produced by a fundamental and its third harmonic

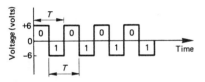

Fig. 7.12 Square data waveform

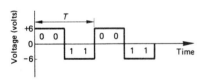

Fig. 7.13 A square data waveform of lower bit rate

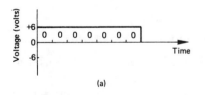

(a)

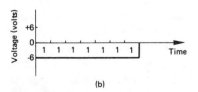

(b)

Fig. 7.14 Data waveforms with zero fundamental frequency

If the fifth harmonic, also with zero phase angle at time $t = 0$, is added to the fundamental and the third harmonic, the resultant waveform is more nearly rectangular. Adding the seventh, ninth, etc. odd harmonics produces an even better approximation to the rectangular waveshape. If all the odd harmonics up to a very high order (theoretically infinity) are all included, a pulse train of square waveform is obtained (see Fig. 7.12).

The number of pulses occurring per second is known as the **pulse repetition frequency** (p.r.f.) and it is equal to the fundamental frequency contained in the pulse waveform. The **periodic time** T of the waveform is the reciprocal of the p.r.f. and is the time interval between the leading edges of consecutive pulses.

A data waveform will only be square when it consists of alternate 0s and 1s as shown in Fig. 7.12. In the periodic time T, one 0 followed by one 1 occur, i.e. two bits. The fundamental frequency of this waveform is equal to $1/T$ and is one-half the number of bits per second. If the data waveform consisted of alternate pairs of 0s and 1s as shown by Fig. 7.13, four bits occur in the periodic time T of the waveform and so the fundamental frequency of the waveform is now equal to one-quarter of the bit rate.

When the data waveform is made up of a number of consecutive 0s or 1s (Fig. 7.14), the data voltage is constant and so the frequency of the waveform is zero hertz.

The data transmitted over a link will include all sorts of combinations of 0s and 1s according to the information content, but the fundamental frequency produced will vary from a minimum of zero hertz to a maximum of one-half the bit rate. The more rapidly the data waveform changes, or in other words the higher the bit rate, the higher will be the frequencies of its components. The minimum bandwidth that must be provided is equal to one-half the bit rate. If only this minimum bandwidth is provided, the data waveform would lose its rectangular shape since none of its harmonics would be transmitted.

EXAMPLE 7.1

Determine the maximum fundamental frequency of a 147 bits/sec data waveform. What other frequencies are present?

Solution

Maximum fundamental frequency $= 147/2 = 73.5$ Hz.

Other frequencies present are

  (i) Third harmonic 220.5 Hz
  (ii) Fifth harmonic 367.5 Hz etc.

If the received current at the end of a cable pair is not allowed sufficient time to reach its maximum value (steady-state) before the signal ends, the waveform of the transmitted signal will not be reproduced. The resultant waveform will not contain all the frequencies predicted from a knowledge of the fundamental frequency of the data waveform. If the time taken for the received current to build up to its steady-state value is less than the time duration of a bit, the receive current waveform will only be affected by the varying loss of the line at different frequencies. If the build-up time of the receive current is greater than the bit length, distortion will occur. Some examples are given in Fig. 7.15. When the time taken for the received current to reach its steady-state value is equal to the bit duration, the receive current waveform is as shown by Fig. 7.15*b*. Clearly the current waveform is not rectangular. If the receive current is able to reach its steady-state value before the trailing edge of the next voltage pulse occurs, its waveshape is approximately rectangular (see Fig. 7.15*c*). Conversely, should the pulse duration be much shorter than the time needed for the receive current to attain its steady state, the current will never reach its steady value and considerable waveform distortion will occur (Fig. 7.15*d*).

Transmission Lines

When the bit rate and the parameters of the line are such that the received current is able to attain its steady-state condition, the received current pulses can be resolved into a fundamental frequency plus a number of harmonically related components, and the response of the line can be determined by the use of transmission line theory.

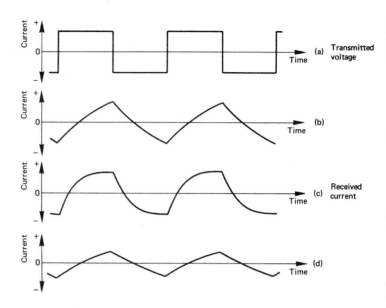

Fig. 7.15 Showing the effect of the build-up time of the received current on the received current waveform; (*b*) build-up time equal to bit duration; (*c*) build-up time less than bit duration; (*d*) build-up time greater than bit duration

The basic principles of operation of transmission lines have been discussed in an earlier volume [TSII] and will be briefly outlined here with particular emphasis on the transmission of pulse waveforms.

The performance of a transmission line is described in terms of its **secondary coefficients,** namely its *characteristic impedance*, its *attenuation coefficient*, its *phase-change coefficient*, and its *velocity of propagation.*

Characteristic Impedance

The **characteristic impedance** of a transmission line is the input impedance of a long length of that line or, alternatively, it is the input impedance of a line that is terminated in the characteristic impedance. A line that is terminated in its characteristic impedance is said to be correctly terminated.

The characteristic impedance of a line depends upon the values of the primary coefficients and also upon the frequency. At zero frequency the characteristic impedance is given by $\sqrt{(R/G)}$ ohms and then falls with increase in frequency until at higher frequencies where $\omega L \gg R$ and $\omega C \gg G$ the impedance becomes constant at $\sqrt{(L/C)}$ ohms. Fig. 7.16 shows how the characteristic impedance of a line varies with frequency.

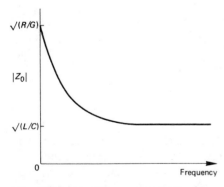

Fig. 7.16 Variation with frequency of the characteristic impedance of a line

Typical values of characteristic impedance for unloaded audio-frequency cable is 600–800 ohms at low audio frequencies falling to about 200–300 ohms at 3000 Hz.

When an audio-frequency cable is *loaded*, i.e. extra inductance is added at 1.828 km intervals along the length of the line [TSII], the value of the characteristic impedance is increased to about 1200 ohms.

Attenuation Coefficient

As a current or voltage is propagated along a line its amplitude is progressively reduced, or attenuated, because of losses in the line. These losses are of two types; first, conductor losses caused by I^2R power dissipation in the series resistance of the conductors; and, secondly, dielectric losses in the insulation separating the conductors. If the current or voltage at the sending end of a line l metres in length is I_S, or V_S, the current, or voltage, at the receiving end is

$$I_R = I_S e^{-\alpha l} \quad \text{or} \quad V_R = V_S e^{-\alpha l}$$

where α is the **attenuation coefficient** of the line in nepers per kilometre. This means that both the current and the voltage decay exponentially as they travel along the line. Usually, the attenuation coefficient is expressed in decibels, i.e.

$$20 \log_{10} I_S/I_R = 20 \log V_S/V_R = \alpha l \tag{7.1}$$

The attenuation of a cable is proportional to the length of the line. Thus if α is 3 dB/km at a particular frequency, the overall loss of a 2 km length of line is 6 dB, of a 4 km length of line is 12 dB, and so on.

Both the resistance and the conductance of a line increase with increase in frequency, and since these are the power-dissipating components the attenuation of the line increases with increase in frequency. Fig. 7.17 shows how the attenuation coefficient of two types of a cable varies with frequency. The attenuation continues to increase at frequencies above those shown in the figure and may be as large as 25 dB/km at 1 MHz. Because of excessive losses like this, wideband telecommunication systems, which employ frequencies of some hundreds of kilohertz upwards, use *coaxial* cable as the transmission medium. The attenuation/frequency characteristic of a coaxial pair is shown in Fig. 7.18. The coaxial pair cannot be used at frequencies below about 60 kHz because the earthed outer conductor would then cease to act as an efficient screen.

EXAMPLE 7.2

A 6 km length of 0.63 mm star-quad cable is used as a data link. Determine the ratio of the attenuations experienced by the funda-

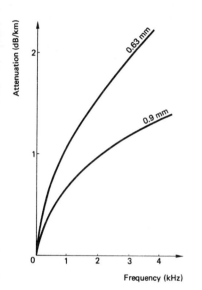

Fig. 7.17 Attenuation/frequency characteristics of audio-frequency star quad cable

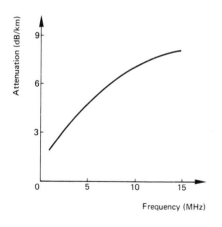

Fig. 7.18 Attenuation/frequency characteristics of a coaxial pair

mental and the third and fifth harmonics of the data waveform when the bit rate is (*a*) 110 bits/sec, (*b*) 1200 bits/sec.

Solution

(*a*) when the bit rate is 110 bits/sec the maximum fundamental frequency is 55 Hz, the third harmonic is 165 Hz, and the fifth harmonic is 265 Hz. From the curve given in Fig. 7.17 it can be seen that the values of the attenuation coefficient of the line at these three frequencies are small and difficult to determine, but clearly they are approximately equal.

(*b*) at 1200 bits/sec the maximum fundamental frequency is 600 Hz, the third harmonic is 1800 Hz, and the fifth harmonic is 3000 Hz. From the curve,

$$\text{at } 600 \text{ Hz } \alpha \simeq 0.8 \text{ dB/km}$$

$$\text{at } 1800 \text{ Hz } \alpha \simeq 1.4 \text{ dB/km}$$

$$\text{at } 3000 \text{ Hz } \alpha \simeq 1.9 \text{ dB/km}$$

Thus the loss of a 6 km length of this cable 4.8 dB at 600 Hz, 8.4 dB at 1800 Hz, and 11.4 dB at 3000 Hz.

4.8 dB is a voltage ratio of 1.74, 8.4 dB is a voltage ratio of 2.63, and 11.4 dB is a voltage ratio of 3.72. Therefore

$$\text{Ratio of attenuation} = 2.63/1.74 = 1.5 \quad \text{(third)}$$

$$\text{Ratio of attenuation} = 3.72/1.74 = 2.14 \quad \text{(fifth)}$$

When cables are connected in cascade their attenuation/frequency characteristics are *additive* as is shown by the following example.

EXAMPLE 7.3

The attenuation/frequency characteristics of four types of audio cable are given by Table 7.1.

Table 7.1

	Attenuation (dB/km)		
	800 Hz	1600 Hz	2000 Hz
Cable A	1.26	1.80	1.95
Cable B	1.13	1.61	1.74
Cable C	0.87	1.23	1.37
Cable D	0.59	0.81	0.39

Plot the overall attenuation/frequency characteristics of links consisting of the cascade connection of

(i) 2 km of cable A, 1 km of cable B, and 2 km of cable C,

(ii) 3 km of cable A, 2 km of cable C, 1.5 km of cable D, and 2.5 km of cable C.

Assume that there are no reflections at any of the cable junctions.

Solution

The required graphs are shown plotted in Fig. 7.19 and are obtained from the data given in Table 7.2.

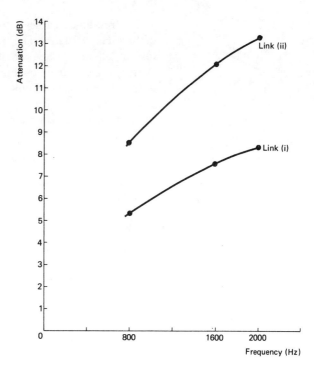

Fig. 7.19

Table 7.2

Link (i)	800 Hz	1600 Hz	2000 Hz
2 km of cable A	2.52	3.60	3.90
1 km of cable B	1.13	1.61	1.74
2 km of cable C	1.74	2.46	2.74
Total loss	5.39	7.67	8.38
Link (ii)			
3 km of cable A	3.78	5.40	5.85
2 km of cable C	1.74	2.46	2.74
1.5 km of cable D	0.89	1.22	1.34
2.5 km of cable C	2.18	3.08	3.43
Total loss	8.59	12.16	13.36

Loading a cable pair decreases the attenuation of the pair at the lower frequencies but produces a rapid increase in the attenuation at frequencies higher than the *cut-off frequency*. This means that a loaded line acts like a low-pass filter. Fig. 7.20 shows the attenuation/frequency characteristic of an

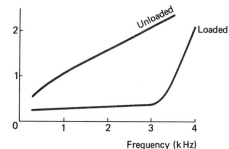

Fig. 7.20 Attenuation/frequency characteristics of loaded and unloaded lines

audio-frequency cable with and without the standard loading of 88 mH/1.828 km. The attenuation of a loaded cable pair is low at all audio-frequencies up to about 3 kHz but it rises sharply at frequencies higher than this. This means that a loaded line may not be able to transmit the higher-frequency components of a data waveform and, particularly if the bit rate is high, waveform distortion will take place.

Phase-change Coefficient

A current or a voltage wave travels along a line with a finite velocity and so the current or voltage at the receiving end of the line lags the current or voltage at the sending end. The total phase lag is equal to the product of the phase-change coefficient of the line and the length of the line in kilometres. The **phase-change coefficient** of a line is the phase lag introduced per kilometre length of line and it is represented by the symbol β.

Phase Velocity

The **phase velocity** of a transmission line is the velocity with which a sinusoidal current or voltage travels along that line. The phase velocity is equal to the product of the signal wavelength and the signal frequency, i.e. $v = \lambda f$ m/s. In a distance of one wavelength along a line, a phase change of 2π radians takes place and so the phase-change coefficient $\beta = 2\pi/\lambda$.

Thus

$$\lambda = 2\pi/\beta$$

and

$$V_p = (2\pi/\beta) \times f = \omega/\beta \text{ m/s} \tag{7.2}$$

The phase velocity of a line will be the same at all frequencies only if the phase-change coefficient β increases with increase in frequency in a linear manner, so that the ratio ω/β is a constant quantity. At audio-frequencies, however, the phase-change/frequency characteristic of a line is not a linear function of frequency (see Fig. 7.21), and so the ratio ω/β is *not* constant. This means that the various component frequencies of a complex wave will arrive at the far end of a line at different times and this will result in waveform distortion.

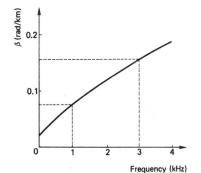

Fig. 7.21 Phase-change coefficient/frequency characteristics of a line

Group Velocity

It is usual to consider the group velocity of a complex wave rather than the phase velocities of its frequency components.

Group velocity is the velocity with which the *envelope* of a complex wave is transmitted.

If a narrow band $\omega_2 - \omega_1$ of frequencies is transmitted over a line, and at these two frequencies the phase-change coefficients of the line are β_2 and β_1 respectively, then the group velocity V_g is given by the equation

$$V_g = (\omega_2 - \omega_1)/(\beta_2 - \beta_1) \text{ m/s} \tag{7.3}$$

The **group delay** of a line is the product of the length of the line and the reciprocal of its group velocity.

EXAMPLE 7.4

A 1200 bits/sec data signal is transmitted over a line whose phase-change coefficient characteristic is shown plotted in Fig. 7.21. The bandwidth of the transmitted signal is so limited that only frequencies up to the fifth harmonic are included. Determine the group velocity of the signal. Find also the group delay if the line is 3 km in length.

Solution
The maximum fundamental frequency of a 1200 bits/sec data waveform is 600 Hz. The third and the fifth harmonics are 1800 and 3000 Hz. Thus the lowest and the highest frequencies to be transmitted are 600 Hz and 3000 Hz respectively. From the graph, the values of the phase-change coefficient at these two frequencies are 0.056×10^{-3} rad/km and 0.155×10^{-3} rad/km respectively.

$$V_g = \frac{2\pi(3000 - 600)}{0.155 - 0.056} = 152 \times 10^3 \text{ km/s}$$

The group delay is

$$\frac{1}{152 \times 10^3} \times 3 = 19.7 \ \mu s$$

The answer to this problem shows that, typically, the group delay per kilometre of an audio-frequency line is of the order of 7 μs per kilometre and this figure is negligibly small. Thus, it is only the high attenuation that stops long unloaded lines being used for digital data signals.

The group delay of a cable is proportional to the length of that cable. Thus if a cable has a group delay of 8 μs per kilometre at a particular frequency the group delay of a 10 km length will be 80 μs.

The group-delay/frequency characteristic of a loaded line depends upon the amount of loading and the length of the line. Fig. 7.22a shows a typical group-delay/frequency characteristics of loaded lines of 16 km length and having loading inductances of 88 mH/1.828 km, 120 mH/1.828 km, and 44 mH/1.81 km. Fig. 7.22b shows group-delay/frequency characteristics of different lengths of 88 mH/1.828 km loaded

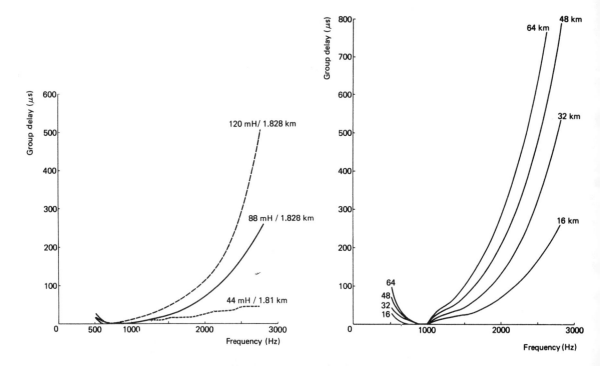

Fig. 7.22 (*a*) Group delay/frequency characteristics of loaded lines in 16 km lengths with different loadings
(*b*) Group delay/frequency characteristics of loaded lines in different lengths of 88 mH/1.828 km loading

line. In these graphs delays of less than 5 μs have been ignored and taken as being equal to zero.

The standard loading used in the United Kingdom telephone network is 88 mH/1.828 km, and most junctions, except those converted for *pulse code modulation* operation, are so loaded. The group delay of a loaded line rises rapidly at frequencies greater than about 1500 Hz and also at frequencies below about 300 Hz. The considerable variation of group delay with frequency means that a data waveform can only be directly transmitted over a line when the bit rate is very low, say 150 bits/sec, unless the line is *very* short.

Group delay distortion in a coaxial cable system is negligibly small.

The Effect of Line Attenuation and Group Delay on Transmitted Pulses and Analogue Signals

The attenuation of an audio-frequency cable increases with increase in frequency and so the various harmonics contained

Fig. 7.23 Effect of line attenuation on a transmitted data waveform

in a data waveform will be attenuated to greater extent than the fundamental frequency. The greater attenuation suffered by the harmonics, particularly the higher-order ones, means that the rectangular waveshape will be lost. This effect is accentuated as the length of the line, and hence its attenuation, is increased, with the result that the pulses become more and more rounded as they travel along a line. This is shown by Fig. 7.23. The higher the bit rate, the higher the fundamental frequency of the data waveform and the shorter the length of line needed to reach the point at which satisfactory reception is impossible. The equipment at the receiving end of the line will respond to each bit at its midpoint and, provided it is able to reliably determine at any instant in time whether a 1 or a 0 is present, reception will be satisfactory.

Because of the effect of line attenuation, direct transmission of data waveforms over unamplified telephone cables is only possible at bit rates of up to 150 bits/sec, and then only for fairly short distances of up to about 4 km.

At these bit rates, the effect of group-delay/frequency distortion is negligibly small.

If the distance between them is *very* short, two computers can be directly connected together without interface equipment and be able to communicate with one another at a high bit rate. The arrangement is shown by Fig. 7.24. Each computer is linked to its outside world by its input/output equipment; this in most cases consists of magnetic and/or paper tape readers/punchers, keyboard printers, and visual display units.

Fig. 7.24 Direct interconnection of two computers

For a reproduced sound signal to appear as natural as possible within the necessary bandwidth restrictions, it is essential that the original amplitude relationships between the fundamental component and the various harmonics are retained. As a signal is propagated along a transmission line it will be attenuated, and this attenuation is greater at the higher frequencies than at the lower. The effect of line attenuation is, therefore, to reduce the amplitudes of the harmonics relative to the fundamental and make the received sound seem unnatural. The effect on short telephone lines is slight and can be tolerated but longer telephone circuits and all-music circuits are normally *equalized* to overcome this problem. A circuit, known as an **equalizer**, is fitted to the receiving end of a line and is adjusted to have an attenuation/frequency characteristic which is the inverse of the attenuation response of the line. The total attenuation of the circuit is then the sum of the attenuations of the line and of the equalizer and is more or less constant. The circuit loss at lower frequencies will have been increased but this can be countered by the use of line amplifiers. Group delay distortion has negligible effect.

The Use of Modems and Amplifiers

The d.c. data signals generated by a computer cannot be transmitted over a link set up via the *public switched telephone network* (p.s.t.n.) and/or over an amplified circuit because the d.c. component will be lost and the data waveform altered. (The d.c. component of a data waveform is the average value.) There are four reasons why the p.s.t.n. cannot transmit the d.c. component of a data waveform:

(1) Transmission bridges in the telephone exchange switching equipment.
(2) Line matching transformers that are used in the audio-frequency junction network to match different cables together and to match line amplifiers to cables.
(3) Line amplifiers incorporate capacitors and/or transformers as coupling components.
(4) Many links will be routed over one or more multichannel telephony systems, or a radio link, and these do not provide a d.c. path.

If the data waveform consists of alternate 1s and 0s, its d.c. component will be zero (Fig. 7.12). At the other extreme when the data waveform consists of a number of consecutive 1s, its d.c. component is −6 V; similarly when consecutive 0s are sent, the d.c. value is +6 V. In either of these cases, the removal of the d.c. component would result in a complete loss of information. Many other data waveforms are also transmitted and, if the d.c. component is lost, the data waveform will be altered.

Suppose, for example, the data waveform is that shown by Fig. 7.25a; the d.c. component or average value of this waveform is $3 \times \frac{6}{5}$ volts (since the 1 pulse cancels out one 0 pulse) or +3.6 V. If this component is removed from the waveform, the resulting waveform will vary from a positive voltage of $+6 - 3.6 = +2.4$ V to a negative voltage of $-6 - 3.6 = -9.6$ V as shown in Fig. 7.25b. The effects of line attenuation and noise will soon ensure that the receiving data terminal will be unable to reliably detect the binary 0 pulses.

The transmission of data over any but the shortest of links necessitates the use of modulation to shift the data waveform to a more convenient part of the bandwidth provided by a

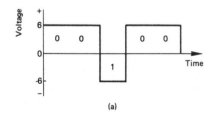

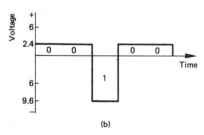

Fig. 7.25 Showing the effect of removing the d.c. component from a data waveform (*a*) with d.c. component, (*b*) without d.c. component

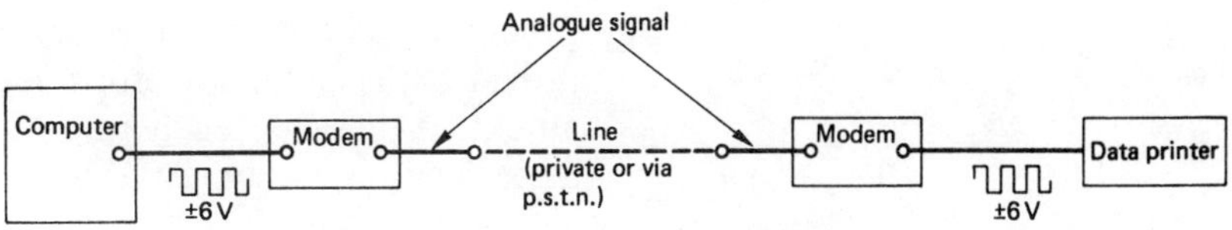

Fig. 7.26 Use of modems for data communication

speech circuit, so that the signal can be transmitted by links set up via the telephone network.

The modulation process is carried out by an equipment known as a **modem** situated at the data source (Fig. 7.26). The data signal is used to modulate a suitable carrier frequency and the modulated (voice-frequency) signal is transmitted over the data link to the distant data terminal. At the distant terminal another modem demodulates or detects the incoming signal to recover the ±6 V data signal. The link may be a point-to-point circuit leased permanently from the telephone authority or it may be a dialled connection set up via the p.s.t.n. In either case the link may be audio-frequency throughout (very possibly amplified) or it may be routed wholly or partly over a multi-channel telephony system or a radio link.

The Use of Regenerators in Digital Systems

As an alternative to the use of modems and amplifiers in an analogue system it is possible to use digital transmission. Many junction cables in the United Kingdom telephone network have been modified by the removal of the loading coils that were originally fitted in order to make the cables suitable for use with **pulse code modulation** systems. This type of system will be discussed in Chapter 10 and it will suffice for the moment to know that the signal information is sent in the form of digital pulse trains. It is possible to transmit the digital data signals over such circuits and to use **pulse regenerators** at regular intervals along the line to reconstitute the pulses. The principle of pulse regeneration is illustrated by Fig. 7.27. The

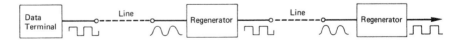

Fig. 7.27 Use of pulse regenerators

pulses transmitted to the line by the data terminal suffer distortion, because of the combined effects of line attenuation and group-delay/frequency distortion and noise, but they are re-created as originally produced by each line pulse regenerator. The data waveform is reconstituted without error provided the pulse waveform has not been degraded to such an extent that the regenerator is unable to reliably determine at each instant in time the presence or absence of a pulse.

Synchronization

When two data terminals are linked together, very likely via modems, and interchanging information, the terminal receiving information must be *synchronized.*

Synchronization is essential so that the receiver will, at all times, sample each incoming bit at the correct instant in time. Otherwise the possibility exists that one, or more, bits may be lost with catastrophic effects on the accuracy of the received data. Suppose, for example, that the three denary numbers 13, 17 and 15 are transmitted using the binary code. The transmitted data is then 011011000101111 (printed in the order of transmission, left-hand side first). If, because of a lack of synchronization, the initial bit is missed, the received data would become 11011000101111 or 27, 2 and either 31 or 30 (depending on whether the next bit to appear is a 1 or a 0).

There are two main methods of synchronization used in data networks known as *anisochronous* and *isochronous.* An **anisochronous** data system, used at low bit rates, inserts synchronizing bits into each character transmitted. Fig. 7.28 shows the example of a 5-bit character produced by a teleprinter; the leading *start* bit turns on the clock in the distant receiver and the trailing *stop* bit turns this clock off. Because the receive clock is turned on and off once per received character, there is insufficient time for any clock inaccuracies to produce a significant error.

In an **isochronous** system, the timing of both the receiver and the transmitter is controlled by a clock in *either* the transmitting data terminal *or* its associated modem. The receiving modem and data terminal derive their timing information from the incoming data itself and/or synchronization bits inserted into the data stream. In one case, for example, the receiver compares the incoming bits with the receive clock, and if need be adjusts the clock to minimize any error. The synchronizing bits are far fewer than those required for an anisochronous system and so the system is more efficient in its transfer of actual data. Isochronous systems work at bit rates of 1200 bits/sec and upwards.

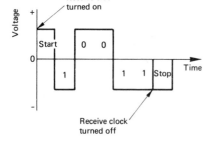

Fig. 7.28 Start-stop synchronization

Exercises

7.1. (*a*) Sketch a block diagram to show how two computers may be directly linked together without interface equipment. Label the input/output equipment used.

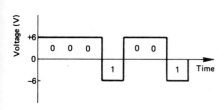

Fig. 7.29

(b) Explain why the system shown in (a) can only be used over very short distances.

7.2. (a) Explain, with the aid of waveform diagrams, the effects of attenuation and group-delay/frequency distortion on a digital data waveform.

(b) Show, with the aid of block diagrams, how the effects of (a) can be overcome using (i) modems and amplifiers, (ii) regenerators.

7.3. (a) Explain, with the aid of suitable sketches, how the waveform of the current at the end of a digital data line link can be constructed using the arrival curve.

(b) Use the arrival curve method to determine the waveform of the received current at the end of a line when the data waveform shown in Fig. 7.29 is applied to the sending end of the line.

7.4. (a) Draw graphs of current and voltage against time at the sending and receiving ends of a digital data line when (i) single-current and (ii) double-current signals are used.

(b) Use your graphs to explain why double-current working gives greater reliability and a faster speed of signalling.

7.5. The attenuation/frequency characteristics of two types of audio-frequency cable are given by the data of Table 7.3. Sketch the overall attenuation/frequency characteristic of 3 km of cable A connected in tandem with 2 km of cable B.

Table 7.3

	Attenuation (dB/km)		
	800 Hz	1600 Hz	2000 Hz
Cable A	1.24	1.78	1.90
Cable B	0.84	1.20	1.33

7.6. (a) Explain what is meant by the terms *attenuation* and *group delay* when applied to a transmission line and briefly state their importance in relation to the transmission of digital data over lines.

(b) A line has a phase-change coefficient of 0.35 rad/km at 4 kHz and of 0.24 rad/km at 2 kHz. Estimate the group velocity of the line at 3 kHz.

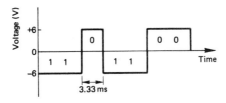

Fig. 7.30

7.7. (a) The digital data waveform shown in Fig. 7.30 has its d.c. component removed. Sketch the resultant waveform.

(b) Why would the d.c. component of a digital data waveform be removed if the signal were transmitted over the p.s.t.n.?

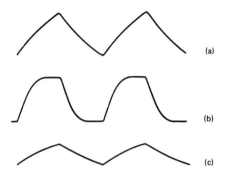

Fig. 7.31

7.8. Explain what is meant by the term *arrival curve* as applied to d.c. data transmission.

Fig. 7.31 shows three received current waveforms which are the result of applying voltage waveforms to three different cable pairs. For each one state whether the time constant of the line is less than, equal to, or greater than the bit length.

7.9. (*a*) What is meant by *loading* a transmission line? What effect does loading have on (i) the characteristic impedance, (ii) the attenuation/frequency characteristic, and (iii) the group-delay/frequency characteristic of a line?

(*b*) What is the effect of (i), (ii) and (iii) on a data waveform and how does this depend upon the bit rate?

7.10. Draw curves to show how the (i) characteristic impedance, (ii) attenuation coefficient, (iii) phase-change coefficient, (iv) phase velocity of propagation of an audio cable vary with frequency.

7.11. A loaded audio cable has the group-delay/frequency characteristic given by the data of Table 7.4. Plot graphs to show the group-delay/frequency characteristic of each length of cable.

Table 7.4

Length in km	Frequency in Hz						
	500	800	1200	1600	2000	2400	
16	30	0	10	30	70	140	Group delay
32	50	10	20	60	140	290	to nearest
48	80	10	30	90	210	430	5 μs

Short Exercises

7.12. A point-to-point data link is routed over audio-frequency cable, a multi-channel telephony system, a high-frequency radio link and then more audio-frequency cable. Draw the block schematic diagram of the circuit.

7.13. List the reasons why a digital data waveform must be converted to analogue form before it can be transmitted over the p.s.t.n.

7.14. Calculate (*a*) the bit rate, (*b*) the highest fundamental frequency of the data waveform shown in Fig. 7.28.

7.15. List the relative merits of single-current and double-current working of a short digital data link.

7.16. Draw the digital data waveform which represents the binary number 011001.

7.17. Draw the circuit of a uni-directional double-current circuit in which the transmitter is a tape reader and the receiver is a tape puncher.

7.18. Draw the circuit of a simplex double-current data circuit.

7.19. If the duration of a bit is 1.67 ms what is the bit rate?

7.20. Explain briefly the difference between the phase and group velocities of a transmission line.

8 Digital Modulation

The bandwidth of the commercial-quality speech circuit is 300–3400 Hz but at frequencies above about 3000 Hz group-delay/frequency distortion increases to such an extent that data transmission becomes difficult. Because of this the highest frequency made available for data transmission is usually 3000 Hz. The public switched telephone network (p.s.t.n.) is unable to transmit signals at or near 0 Hz because of line matching transformers, transmission bridges in the telephone exchange equipment, and the widespread use of amplified circuits, both audio and multi-channel. The direct transmission of a data waveform over a telephone network is only possible for low bit rates up to 150 bits/sec over non-amplified lines, or for higher bit rates over *very* short lines. For all other data links the data waveform must be applied to a modem in which a carrier frequency can be modulated in amplitude, frequency, or phase to produce an analogue signal which can be transmitted over the telephone network.

Modems

A **modem** is an equipment that is fitted to each end of a data circuit to change the serial digital signals produced by the data terminal into voice-frequency signals suitable for transmission over the telephone line network. A modem also converts voice-frequency signals received from line into serial binary data signals which are then passed to the data equipment. A modem may be able to operate in one or more of the *simplex*, *half-duplex* and *full-duplex* modes.

Essentially, a modem has two parts: a transmitter and a receiver. The basic block diagram of the transmitting section of a modem is shown by Fig. 8.1. The encoder is needed when *four-phase*, or *four-level*, modulation is used to group the incoming bits into *dibits* (see p. 107). The carrier frequency is applied to the modulator and is modulated by the data signal to produce the corresponding voice-frequency signal. The bandwidth of the modulated data waveform is limited by the bandpass filter to restrict the transmitted signal to the bandwidth made available by the line. The band-limited signal is amplified before it is transmitted into the two-wire or four-wire line. The amplifier also ensures that the modem is impedance-matched to the line over the range of frequencies to be transmitted.

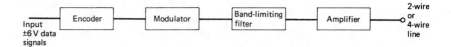

Fig. 8.1 Transmitting section of a modem

Fig. 8.2 Receiving section of a modem

Fig. 8.2 shows the basic block diagram of the receiver section of a modem. The incoming voice-frequency signals are filtered to remove unwanted noise and distortion components lying outside the wanted frequency band. The incoming signals are then amplified by the amplifier whose gain is controlled by an automatic gain control circuit to ensure that the input voltage to the demodulator remains more or less constant at all times. The demodulator extracts the information content of the modulated waveform to produce a ±6 V signal; if the output signal is in dibit form it is then decoded into the wanted serial form.

Not all of the frequency spectrum of a commercial-quality speech circuit can be made available for the transmission of data since it is necessary to avoid those frequencies which are used for signalling and supervisory purposes. For this reason the available frequency spectrum is restricted to 300–500 Hz and 900–2100 Hz, i.e. a bandwidth of 1400 Hz. (2280 Hz is the main signalling frequency used in modern systems.) Low-speed data links of up to 1200 bits/sec can be operated so that each bit in the data waveform is represented by a single change in the modulated parameter. Higher-speed modems, operating at 2400 bits/sec or more, must generally encode the data waveform before the modulation process is carried out. One method used is to pair bits together to form **dibits**, each dibit being used to produce a single change in the modulated carrier. The four possible dibits are 00, 01, 10 and 11. Since changes in the modulated carrier will now occur only *half* as often as when single bits are used, the *effective* bit rate on the transmission medium is reduced by half. The bit grouping process can be carried out still further so that a single change in the modulated carrier occurs for every *four* bits of information sent; the effective transmitted bit rate will then be reduced fourfold but at the expense of increased circuit complexity and hence increased cost.

The digital modulation methods that are commonly used for data communication are *frequency shift modulation or keying, differential phase modulation*, and *vestigial sideband amplitude modulation.*

For bit rates up to and including 1200 bits/sec frequency shift modulation is the standard method of modulation used. Frequency shift modulation cannot be used at higher bit rates because the bandwidth needed becomes too great. For bit rates of 2400 bits/sec and higher, differential phase modulation is employed. At the very high bit rate of 48 kilobits/sec, vestigial sideband amplitude modulation is used.

Frequency Shift Modulation

When a sinusoidal carrier wave is frequency modulated, its frequency is made to vary in accordance with the characteristics of the modulating signal. The amount by which the carrier frequency is deviated from its nominal value is *proportional* to the amplitude of the modulating signal, and the number of times per second the carrier frequency is deviated is *equal* to the modulating frequency.

When the modulating signal is a data waveform, the amplitude of the modulating signal is ±6 V and its maximum fundamental frequency is one-half of the bit rate. This means that the carrier frequency is *always* deviated to either one of two different frequencies; the nominal carrier frequency is the arithmetic mean of these two frequencies but it is never actually present since the modulating signal is never at 0 V. The higher of the two frequencies is used to represent binary 0 while the lower frequency represents binary 1. The choice of the two frequencies is a compromise between the bandwidth occupied by the modulated signal and the wanted signal-to-noise ratio at the receiver, since both of these parameters increase with increase in the frequency deviation employed.

Table 8.1

Bit rate (bits/sec)	Frequencies (Hz) Binary 0	Binary 1		Nominal carrier frequency (Hz)
up to 300	1180 1850	980 1650	(different directions of transmission)	1080 1750
600	1700	1300		1500
1200	2100	1300		1700
75/150 (supervisory channel)	450	390		420

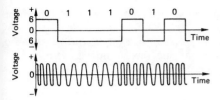

Fig. 8.3 Frequency-shift waveform

For lower data rates it is possible to accommodate two channels and thus provide duplex transmission on a two-wire speech circuit. The frequencies used to represent the bits 1 and 0 for the standard bit rates are given by Table 8.1. In each case the nominal carrier frequency is the arithmetic mean of the two frequencies.

An example of a frequency-shift waveform is given by Fig. 8.3. When the data waveform is at +6 V (binary 0), the transmitted frequency is the higher of the two frequencies and, similarly, the lower frequency is transmitted to represent binary 1.

Modulation Index and Deviation Ratio

The modulation index m_f of a frequency-modulated wave is the ratio of the frequency deviation to the modulation frequency, i.e.

$$m_f = \frac{\text{Frequency deviation}}{\text{Modulation frequency}} \tag{8.1}$$

The deviation ratio D of a frequency-modulated wave is the particular case of the modulation index when *both* the frequency deviation *and* the modulating frequency are at their maximum values, i.e.

$$D = \frac{\text{Maximum frequency deviation}}{\text{Maximum modulating frequency}} \tag{8.2}$$

In the case of frequency shift modulation, the frequency deviation is fixed since the amplitude of the modulating signal voltage—the data waveform—is ±6 V. The maximum modulating frequency exists when the data waveform consists of alternate 1s and 0s and it is then equal to one-half of the bit rate. Thus the deviation ratio of a frequency shift system is given by

$$D = \frac{\text{Frequency deviation}}{(\text{Bit rate})/2} \tag{8.3}$$

EXAMPLE 8.1

Calculate the deviation ratio for a 1200 bits/sec frequency-shift data system.

Solution
For a 1200 bits/sec data system the two frequencies used are 1300 Hz and 2100 Hz and the nominal carrier frequency is

$$(1300 + 2100)/2 = 1700 \text{ Hz}$$

Thus, the frequency deviation is 400 Hz. Hence

Deviation ratio = 400/600 = 0.67

In this calculation of the deviation ratio the harmonics of the data waveform have not been considered.

The Frequency Spectrum of a Frequency-shift Waveform

When a sinusoidal carrier wave of frequency f_c is frequency modulated by a sinusoidal signal of frequency f_m, the modulated wave may contain components at a number of different frequencies. These frequencies are the carrier frequency and a number of sidefrequencies positioned either side of the carrier. The sidefrequencies are spaced apart at frequency intervals equal to the modulating frequency.

The first-order sidefrequencies are $f_c \pm f_m$.
The second-order sidefrequencies are $f_c \pm 2f_m$.
The third-order sidefrequencies are $f_c \pm 3f_m$
and so on.

The amplitudes of the various frequency components depend upon the value of the deviation ratio [RSIII], and even the carrier component may be zero at some particular values of D (2.405, 5.52, 8.654),
The bandwidth needed for the transmission of a frequency-modulated wave is given by

$$\text{Bandwidth} = 2(f_d + f_m) \tag{8.4}$$

where f_d is the maximum frequency deviation and f_m is the maximum modulating frequency.
This expression assumes that all sidefrequencies whose amplitude is greater than 10% of the unmodulated carrier amplitude must be transmitted in order that distortion of the received waveform should be negligibly small. For a frequency-shift waveform the bandwidth expression can be rewritten as

$$\text{Bandwidth} = 2(f_d + \text{bit rate}/2) \tag{8.5}$$

Suppose a 1200 bits/sec frequency-shift waveform is to be transmitted. From equation (8.5) the bandwidth needed is

$$\text{Bandwidth} = 2(400 + 1200/2)$$

$$= 2000 \text{ Hz}.$$

This means that the bandwidth needed to transmit a 1200 bits/sec frequency-shift system is greater than the bandwidth that is available (900–2100 Hz or 1200 Hz) when a data system is set up over the p.s.t.n. Fortunately it is not necessary that the waveform of the signal arriving at the far end of a link is undistorted. It is sufficient that the receiving equipment is able to determine at any instant whether a binary 1 or binary 0 bit is being received. This means that it is *not* necessary for all the significant sidefrequencies to be transmitted. Since the deviation ratio of a frequency-shift waveform is approximately unity, most of the energy content of the waveform is concentrated in the carrier and the first-order sidefrequency components. Hence, *only* these components need to be transmitted.

The advantage of only transmitting the first-order sidefrequencies is the considerable reduction in the occupied bandwidth that is obtained. The necessary bandwidth is only *twice* the maximum modulating frequency or

$$\text{Bandwidth} = \text{Bit rate} \qquad (8.7)$$

In the previous example of a 1200 bits/sec frequency-shift system the necessary bandwidth is now only 1200 Hz as opposed to the 2000 Hz previously calculated. This narrower bandwidth can be accommodated in the frequency spectrum made available by the p.s.t.n. although it should be realised that 1200 bits/sec represents the maximum bit rate that can be transmitted.

The basic block diagram of a **frequency shift keying modulator** is shown by Fig. 8.4. The data waveform to be transmitted is band-limited by the input bandpass filter which has a cut-off frequency [TSII] of 1300 Hz. The waveform applied to the voltage-controlled astable multivibrator [EIII] consists of the fundamental frequency only of a 1200 bits/sec signal or the fundamental plus third harmonic of a 600 bits/sec signal. The voltage-controlled multivibrator is switched between two states: one in which it oscillates at 1300 Hz and one in which it oscillates at either 1700 Hz or 2100 Hz (depending on the bit rate). The square output waveform of the multivibrator is applied to a bandpass filter whose bandwidth is sufficiently narrow to ensure that only the fundamental frequency of either 1300 Hz or 1700 Hz (or 2100 Hz) is passed. The output of the filter is thus the wanted frequency-shift waveform. Another type of f.s.k. modulator consists essentially of an *LC* oscillator whose tuned circuits values are switched from one set of values to another by the ±6 V data signals.

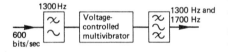

Fig. 8.4 F.S.K. modulator

Phase Modulation

Frequency shift modulation is only available for bit rates up to 1200 bits/sec since the next higher standard bit rate of 2400 bits/sec would need a bandwidth of 2400 Hz which cannot be provided by the p.s.t.n. For 2400 bits/sec systems therefore, *phase modulation* is used.

When a sinusoidal carrier is phase modulated, its instantaneous phase is made to vary in accordance with the characteristics of the modulating signal. The magnitude of the phase deviation is *proportional* to the modulating signal voltage and the number of times per second the phase is deviated is *equal* to the modulating frequency. In a data system the modulating signal voltage is ±6 V and so the phase deviation obtained is fixed.

Modulating the phase of the carrier will at the same time vary the instantaneous carrier frequency since angular velocity ($\omega = 2\pi \times$ carrier frequency) is the rate of change of phase.

Modulation Index and the Frequencies Contained in a Phase-modulated Waveform

The modulation index of a phase-modulated waveform is, as with frequency modulation, the maximum phase deviation of the carrier frequency produced by the modulating signal. However, the value of the modulation index depends only upon the modulating signal voltage and is quite independent of the modulating frequency.

The modulation process generates a number of sidefrequencies spaced symmetrically either side of the carrier frequency, and the frequency spectrum is exactly the same as that of a frequency-modulated wave having the same numerical value of modulation index.

An example of the use of phase modulation is given by Fig. 8.5. The phase of the carrier is shifted by 180° each time the leading edge of a 0 bit occurs. The phase of the carrier is not altered by a 1 bit. In practice, this form of phase modulation is rarely employed and instead **differential phase modulation** is commonly employed. This version of phase modulation uses *changes* in phase, rather than phase itself, to indicate the binary digits 1 and 0.

Four phase changes are used, each of which represents a dibit. Two systems are given in Table 8.2; the upper row of phase changes was the United Kingdom standard until recently and is still used in the U.K. data network. The lower row of phase changes is the system recommended by the C.C.I.T.T.† which has been supplied for new equipment for a few years now and has now been adopted as the standard U.K. system.

The use of dibits reduces the bandwidth that must be made available since the maximum fundamental frequency of a dibit waveform is $\frac{1}{4} \times$ bit-rate.

When dibits are used, the equipment at the receiving end of the system must be synchronized with the transmitting equip-

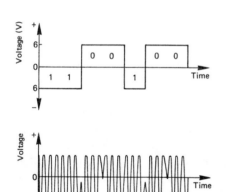

Fig. 8.5 Phase modulated waveform

---

†C.C.I.T.T. Consultative Committee for Telegraphy and Telephony.

Table 8.2

Dibit	0 0	0 1	1 1	1 0
Phase change	0°	+90°	+180°	+270°
Phase change	+45°	+135°	+225°	+315°

ment for decoding to take place correctly. The necessary synchronization is developed from the changes in phase of the received signal. In the earlier system, errors could be caused by loss of synchronization resulting from a long string of 0s and thus no changes in phase at the receiver. The newer C.C.I.T.T. system has the advantage that the dibit 00 is denoted by a 45° phase shift and not by 0° phase change.

Suppose for example that the data signal 10100111001101 is to be transmitted (in the order printed). The bits are grouped together in the decoder section of the modem to form the dibits 10, 10, 01, 11, 00, 11 and 01. These dibits are then signalled to line by changing the phase of the carrier by, in turn, +225°, +225°, +135°, +315°, +45°, +315° and +135°. The carrier frequency used is 1800 Hz since at this frequency the group-delay/frequency distortion of a line is small and the frequency is approximately at the middle of the available frequency spectrum.

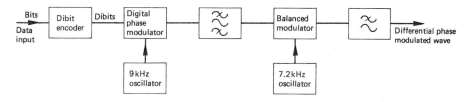

Fig. 8.6 Differential phase modulator

The basic block diagram of a differential phase modulator is shown in Fig. 8.6. The incoming bit stream is encoded into dibits and then passed onto a digital phase modulator which is fed with a 9 kHz carrier. The output of the modulator is band-limited and then applied to an amplitude-modulation balanced modulator together with the output of a 7.2 kHz oscillator. The balanced modulator produces the sum and the difference of its input frequencies and the 7.2 kHz carrier, and the actual frequencies generated for each dibit are given by Table 8.3. The top row of frequencies are those used in the older system and the bottom row gives the frequencies used in the newer C.C.I.T.T. system. It should be noted that with one exception (750 Hz for dibit 00) the frequencies transmitted to line in the C.C.I.T.T. system are equal to the older system frequencies *plus* 150 Hz.

Table 8.3

Dibit	0 0	0 1	1 0	1 1
Frequencies transmitted (Hz)	1800	900 2100	1500 2700	1200 2400
Frequencies transmitted (Hz)	750 1950	1050 2250	1650 2850	1350 2550

The bandwidth needed to accommodate the 2400 bits/sec phase-modulated system is either $2700 - 900 = 1800$ Hz, or it is $2850 - 750 = 2100$ Hz. In both cases the required bandwidth is less than would be needed for a frequency-shift system using the same bit rate, but it still encroaches into the part of the p.s.t.n. spectrum occupied by signalling tones. Also, a p.s.t.n. connection is likely to have a poor group-delay/frequency characteristic at higher frequencies and usually this is much more of a problem than the presence of signalling tones. Most 2400 bits/sec systems are therefore operated over privately leased circuits which will probably not have associated in-band signalling equipment and which can be adjusted to have a good group delay characteristic. When a 2400 bits/sec system is worked over the p.s.t.n., it may or may not work satisfactorily depending on the characteristics of the actual link set up. The actual bandwidth used for signalling frequencies is 2130–2430 Hz and, in practice, data signals *can* fall within this band provided there are also signals in the band 900–2130 Hz whose amplitude is not less than a set figure.

It is possible to use more than four different phase change values to represent groups of bits and in so doing obtain even greater economy in the use of the available frequency spectrum. If eight phases are used, the bits can be grouped in threes such as 000, 001, etc. and then the effective bit rate for transmission over a line is reduced threefold, or put the other way, a 1200 baud modulation rate can accommodate a 3600 bits/sec data waveform. Similarly, if the bits are grouped in fours such as 0000, 0001, etc. there are 16 different combinations, and consequently 16 different phase changes are needed. In this case a 1200 baud line will be able to transmit a $4 \times 1200 = 4800$ bits/sec signal. Clearly, increasing the number of phases used allows a more efficient usage of the line but this advantage is offset by an increase in the complexity of the receiving equipment.

Amplitude Modulation

When a carrier wave is amplitude modulated, its amplitude is caused to vary in accordance with the characteristics of the

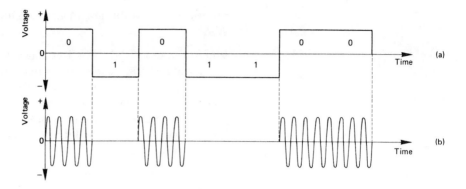

Fig. 8.7 Amplitude modulated waveform

modulating signal. The amplitude of the modulated carrier will vary, with the same waveform as the modulating signal, between a maximum value, equal to the sum of the carrier voltage and the modulating signal voltage, and a minimum value, equal to the carrier voltage minus the modulating signal voltage. The outline of the modulated carrier voltage is known as the *modulation envelope*. Fig. 8.7*b* shows the waveform of a sinusoidal carrier amplitude modulated by the data waveform shown in Fig. 8.7*a*. The carrier wave is effectively switched on and off by the data signal.

Depth of Modulation

The modulation factor m of an amplitude modulated wave is defined by the expression

$$m = \frac{\text{Maximum voltage} - \text{Minimum voltage}}{\text{Maximum voltage} + \text{Minimum voltage}} \tag{8.8}$$

When the modulation factor is expressed as a percentage, it is generally called the **modulation depth** or the **depth of modulation**.

When the modulating signal is a $\pm 6\,\text{V}$ data waveform, the modulation factor becomes

$$m = \frac{(V_c + 6) - (V_c - 6)}{(V_c + 6) + (V_c - 6)} = 6/V_c \tag{8.9}$$

where V_c is the peak value of the carrier voltage.

Thus the modulation factor is equal to the ratio of the data and the carrier voltages.

Frequencies Contained in an Amplitude-modulated Wave

When a sinusoidal carrier frequency is amplitude modulated by a data signal, each component frequency of the data waveform will give rise to *two* sidefrequencies in the modulated waveform. If, for example, the data waveform consists of a series of alternate 0 and 1 bits, its fundamental frequency is equal to one-half of the bit rate and the modulated wave will contain components at

    (i) the carrier frequency f_c
   (ii) the upper sidefrequency $f_c + (\text{bit rate}/2)$
  (iii) the lower sidefrequency $f_c - (\text{bit rate}/2)$
  (iv) upper and lower sidefrequencies $f_c \pm 3 \times (\text{bit rate}/2)$
      etc.

The band of sidefrequencies below the carrier is known as the lower sideband and the band of sidefrequencies above the carrier frequency is called the upper sideband.

EXAMPLE 8.2

A 2400 bits/sec data waveform amplitude modulates a 1800 Hz carrier wave. Calculate the necessary bandwidth for the transmission of this signal if (i) only the maximum fundamental frequency, (ii) the maximum fundamental frequency and its third harmonic need be transmitted.

Solution
(i) The maximum fundamental frequency component is 2400/2 or 1200 Hz.

$$\text{Bandwidth} = 2f_{m(max)} = 2400 \text{ Hz}$$

(ii) Third harmonic of 1200 is 3600 Hz.

$$\text{Bandwidth} = 7200 \text{ Hz}$$

This example shows that the bandwidth occupied by a double-sideband (d.s.b.) amplitude-modulated data signal is wider than that demanded by the differential phase modulation system.

Greater economy in bandwidth can be realized if multi-level amplitude-modulation is used. Four levels of amplitude can be used to represent the four dibits (Fig. 8.8). The dibit waveform amplitude-modulates the carrier wave in the usual way to produce the waveform shown. The bandwidth occupied by this waveform is halved but the receiving equipment is now more complex and hence expensive.

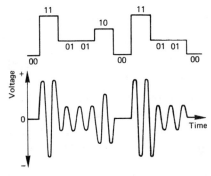

Fig. 8.8 Multi-level amplitude modulation

Single-sideband Operation

The information represented by the modulating signal is con-

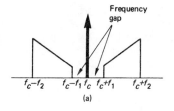

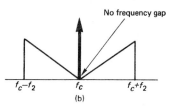

Fig. 8.9 Spectrum diagrams of amplitude-modulated waveforms: (a) when the minimum modulation frequency is f_1; (b) when the minimum modulation frequency is 0 Hz

tained in both the upper and the lower sidebands and this means that it is not necessary to transmit both sidebands. Either sideband can be suppressed at the transmitting end of a system without any loss of information. Furthermore, the carrier component is of constant amplitude and frequency and does not carry any information and so it also need not be transmitted. Single-sideband (s.s.b.) operation of a communication system offers a number of advantages over d.s.b. working but the most important is that the occupied bandwidth is halved.

In line and radio systems it is possible to suppress the carrier and the unwanted sideband by using the combination of a balanced modulator and a band-pass filter. This practice is possible because the audio bandwidth to be transmitted does not go down to 0 Hz and so a frequency gap exists between the carrier and the edge of the wanted sideband. This is illustrated by 8.9a. The fundamental frequency of a data waveform varies according to its information content but may be zero when consecutive 0 or 1 bits are transmitted. This means that the two sidebands produced by the modulation process will not have a frequency gap between them and the carrier (Fig. 8.9b).

A filter needs some frequency gap between wanted and unwanted frequencies so that its attenuation can build up [TSII] and for this reason straightforward s.s.b. operation is not possible for a data system. S.s.b. operation is possible if complex terminal equipment is used at both ends of a link but such a system is not used in the U.K.

Another version of amplitude modulation is known as **vestigial sideband** (v.s.b.) **modulation**. In a v.s.b. system, one sideband is transmitted along with a reduced level carrier and a part, or *vestige*, of the other sideband. Vestigial sideband a.m. is used in the U.K. network for very high speed data links operating at 48 kilobits/sec. Such high speed links are used to directly connect two digital computers together. The maximum fundamental frequency of a 48 kilobits/sec data waveform is 24 kHz which is far in excess of the bandwidth provided by an audio speech circuit. In any case, in practice frequencies up to 36 kHz are transmitted to line.

If d.s.b. a.m. were to be used, a bandwidth of 72 kHz would be necessary but the use of v.s.b. reduces this to about 44 kHz. A convenient transmission medium which is well able to transmit such a bandwidth is available in the trunk telephone network, namely the line equipment of the 12-channel telephony group [TSII]. This medium provides a bandwidth of 60–108 kHz. To position the data signal in this bandwidth the signal amplitude modulates a 100 kHz carrier frequency to produce the v.s.b. a.m. waveform shown in Fig. 8.10.

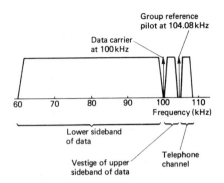

Fig. 8.10 V.S.B. waveform (from the *Post Office Electrical Engineer's Journal*)

The lower sideband of modulation occupies the band 60–100 kHz and the vestige of the upper sideband that is retained occupies the band 100–104 kHz. This vestige is sufficiently wide to simplify the design of the v.s.b. filters. The carrier frequency is only transmitted at a low level so that it can be used at the receiving end of the system to lock a locally-generated carrier at the correct frequency and phase so it can be used for the demodulation process. A telephone channel can be provided in the band 104–108 kHz but this facility is no longer provided by the British Post Office.

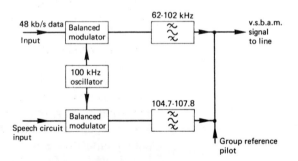

Fig. 8.11 V.S.B. modulator

The basic block diagram of a v.s.b. modulator is shown by Fig. 8.11. The 48 kilobits/sec data signal is applied to a balanced modulator to produce upper and lower sidebands and a low-level carrier. The modulating signal actually occupies the band 2–36 kHz and is derived from another modem. The bandpass filter following the modulator has a nominal pass band of 62–102 kHz but passes frequencies 2 kHz either side of this band with increasing loss. This is the reason for the sloping edges of the sideband characteristics shown in Fig. 8.10.

Vestigial sideband amplitude modulation is used because of the bandwidth economy it provides; for example none of the alternative modulation methods previously described could give a 48 kilobits/sec system in a 60–108 kHz bandwidth. To greatly reduce the probability of the system causing interference with adjacent 60–108 kHz links carrying multi-channel telephony signals when repetitive data patterns are transmitted, an encoder is fitted at the point at which the data originates to code the signal and break up any repetitive patterns and to spread the power content of the data signal over the available bandwidth. A similar equipment must be fitted at the far end of the data link to decode the incoming signal to obtain the original data waveform.

The Effects of Line Attenuation and Group-delay/Frequency Distortion on Modulated Data Signals

When modulated data signals are transmitted over a line, attenuation and group-delay/frequency distortion may distort the received signal to such an extent that the receive modem may not be able to interpret the incoming signals correctly. The error rate will then rise. The effects of the line characteristics will be explained by considering frequency shift modulation.

When frequency shift modulation is used, the higher frequency transmitted will be attenuated to a greater extent than the lower frequency and hence the 0 bits may become of so low a level that it becomes difficult to determine when a 0 bit has been received. This effect is much more prevalent for 1200 bits/sec systems than for 600 bits/sec systems because the former system uses the higher frequency of 2100 Hz to represent a 0 bit. Thus the effect of line attenuation is to reduce the maximum data transmission rate. The effect of group-delay/frequency distortion is to delay the 2100 (or 1700) Hz pulses to a greater extent than 1300 Hz pulses, so that the higher frequency pulses arrive at the end of the link at incorrect instants in time (relative to the lower frequency bits). The effect of this is to make the pulses overlap so that binary 1 appears longer and binary 0 appears shorter than they should. In extreme cases the receive modem may not be able to detect a 0 bit received in between two 1 bits.

Similar but smaller effects occur when differential phase modulation is used because the changes in phase are signalled to line by various pairs of frequencies.

Exercises

8.1. Explain the basic principle of operation of two-phase and four-phase differential phase modulation systems.

The data signal 1110 01 00 10 is to be transmitted using
(i) 4-phase differential modulation with 00 indicated by 0° phase change,
(ii) 4-phase differential phase modulation with 00 indicated by +45° phase shift,
(iii) 2-phase differential phase modulation with 00 represented by +90° phase shift.

For each system write down the carrier phase changes that are successively transmitted to line.

8.2. (a) Outline a method by which data may be transmitted over an audio-frequency circuit at a bit rate of 600 bits/sec.
(b) How can errors in the received data be detected?

8.3. Explain, with the aid of waveform sketches, what is meant by frequency shift modulation. A 1500 Hz carrier wave is frequency shift modulated by a 600 bits/sec data waveform.
(a) What is the maximum fundamental frequency of the data waveform?
(b) What bandwidth is required for the transmission of the frequency-shift waveform if (i) the fundamental component only is transmitted, (ii) the third harmonic is also transmitted?
(c) Why is it possible to transmit only the first-order sidefrequencies and what is the advantage gained by so doing?

8.4. What is meant by the term *dibit* used in data communications? How can dibits be represented by (a) changes in the phase angle of a carrier wave, (b) changes in the amplitude of a carrier wave? State the advantage which is gained by the use of multi-phase or multi-level transmission.

8.5. Draw the waveform of a carrier wave that has been frequency-shift modulated by the data signal 1011001 at a bit rate of 1200 bits/sec. Label the diagram with the frequencies that are transmitted and state the nominal carrier frequency. Which data waveform will have the highest fundamental frequency and what is then the occupied bandwidth?

8.6. What is meant by frequency-shift modulation when applied to data transmission? Explain how the performance of a frequency-shift system can be adversely affected by line attenuation and group-delay/frequency distortion. Why is a 1200 bits/sec system more likely to be affected than a 600 bits/sec system?

8.7. Draw the block diagram of a modem, showing both transmitting and receiving sections, and briefly explain its operation. Discuss the factors that influence the choice of carrier frequency.

8.8. What is the bandwidth of a commercial-quality speech circuit? Why cannot all of this bandwidth be employed for data transmission? State what bandwidth is available (a) for a frequency-shift system, (b) for a differential phase system. Why is amplitude modulation not generally used for data transmission?

8.9. How does frequency shift modulation differ from frequency modulation? Illustrate your answer with waveform sketches. Calculate the deviation ratio for a 600 bits/sec frequency-shift data system. What bandwidth is needed for such a system?

8.10. What is meant by *vestigial sideband amplitude modulation* and why is it used in some data communication systems? Give brief details of the 48 kilobits/sec v.s.b. system used in the United Kingdom and draw the spectrum diagram of this system.

Short Exercises

8.11. Define *bit rate*. How is the bandwidth needed to transmit a data signal related to the bit rate?

8.12. A 600 bits/sec frequency-shift data system is transmitted over a line. What is the bandwidth that must be provided?

8.13. What is the bandwidth of a speech channel? Why is this bandwidth not usually available for data transmission over the p.s.t.n.?

8.14. What is the advantage of differential phase modulation over frequency shift modulation for the transmission of data?

8.15. With the aid of frequency spectrum diagrams explain briefly what is meant by a vestigial sideband amplitude-modulated wave.

8.16. A 1200 baud data channel is to be used to transmit differential phase modulated data signals. What bit rate can be transmitted if (i) 4 phases are used, (ii) 8 phases are used?

8.17. What is meant by the term dibit when applied to a data system? Divide the signal 00110101 into dibits.

8.18. Calculate the nominal carrier frequency of a frequency-shift data system if the bit rate is (i) 600 bits/sec, (ii) 1200 bits/sec.

8.19. Explain by means of a frequency spectrum graph what is meant by an amplitude-modulated vestigial-sideband signal.

8.20. List the advantages and disadvantages of v.s.b. amplitude modulation.

9 Data Links

Modern industry, commerce and government rely to a considerable extent on the digital computer to carry out a wide variety of tasks. Amongst these are the storage of engineering, scientific, and medical records, the calculation of wages, salaries, bills and accounts, and the printing of payslips, accounts and so on. The high-street banks make use of computers to maintain details of customers accounts, of standing orders, direct debits, etc., while airlines and package holiday firms are able to operate booking systems that provide an immediate confirmation of vacancies and bookings.

The cost of a digital computer with a large storage capacity is high and so it is not economic for an organization to instal a computer at all the points in its offices and factories where a computing facility would be of use. It is only economic for a computer to be installed at one, or perhaps two, points in the organization's network of offices, etc. For the branch offices to have on-line access to computing facilities it is necessary for them to be connected to the computer centre by means of *data links*.

Many smaller businesses are unable to economically justify the cost of purchasing and operating a digital computer and yet have a need for computing facilities. To meet this demand *computer bureaux* have been set up which rent computer services to their customers on a time-sharing basis.

The Use of Privately Leased Circuits and the P.S.T.N.

A data link that connects a *data terminal* to a remote digital computer may be leased from the telephone administration or it may be temporarily set up by dialling a connection via the *public switched telephone network* (p.s.t.n.). The choice between leasing a private circuit and using the p.s.t.n. must be made by the careful consideration of factors such as the cost, availability, speed of working, and transmission performance. Private circuits may transmit d.c. ±6 V or ±80 V signals or may use modems to convert the data signals into voice-frequency signals. The maximum length of a ±6 V data link is short but when ±80 V signals are used almost any distance can be covered albeit at only a low bit rate (i.e. up to 150 bits/sec). Voice-frequency data circuits can also be of any length and may be routed, wholly or partly, over loaded or unloaded audio-frequency cable, over multi-channel telephony systems, or over a radio link. (A multi-channel system may itself be routed, wholly or partly, over a radio link.)

A point-to-point privately leased circuit provides exclusive use of the transmission path and so its parameters, such as its attenuation/frequency and group-delay/frequency characteristics, can be equalized and adjusted to give the optimum performance which enables high bit rates to be reliably transmitted. A p.s.t.n. connection is established by dialling the telephone number of the distant data terminal and the route used for a particular call is a random choice from a large number of different possible routings, including various combinations of audio-frequency cable and multi-channel system. Because of this the overall attenuation and group-delay/frequency characteristics of a p.s.t.n. connection are not predictable and so they cannot be adjusted to give the best transmission performance. Transmission at 2400 bits/sec and 4800 bits/sec is available over the p.s.t.n. but the results cannot be guaranteed; some of the dialled connections may not be adequate and the error rate may be too high.

The bandwidth provided by a leased circuit is 300–3000 Hz, the lines being adjusted to give a good transmission performance over this bandwidth. At higher frequencies group-delay/frequency distortion increases rapidly. The bandwidth made available by the p.s.t.n. is more restricted, nominally 300–500 Hz and 900–2100 Hz, because of the need to avoid the frequencies used for in-band signalling systems on trunk routes. Signalling tones may also be present on some leased circuits.

A p.s.t.n. connection is noisier than a leased circuit because switching equipment in telephone exchanges is a source of considerable noise and interference. The effect of noise and interference voltages is to reduce the maximum bit rate that can be employed for a given error rate.

The use of a privately-leased circuit for data transmission has the following advantages over the use of the p.s.t.n.:

(1) Exclusive use of the circuit is obtained. Thus time is not wasted setting up calls and the performance of the circuit remains stable.

(2) The link can be adjusted to have the optimum performance and it is less affected by noise and interference. As a result higher speed and more reliable transmissions are possible.

(3) Full-duplex operation is available at higher bit rates and higher data "throughput" is possible by eliminating turn-round time. (With half-duplex operation the time taken for the modem to switch from reception to transmission and vice versa—known as the **turn-round time**— is some tens of milliseconds.)

On the other hand the cost of permanently leasing a line may be relatively high and it can only be economically justified if there is sufficient data traffic on the line or the particular terminal application necessitates a permanent connection, e.g. a cash dispensing terminal in a bank which checks a customer's account before releasing the cash. If the data communication requirements involve occasional contact with a large number of locations, and the majority of the connections are of fairly short duration, the use of the p.s.t.n. is probably best. On the other hand if long duration connections between a few branch offices and the computer centre is planned for, leased links will probably be chosen. In practice, most private data networks consist of a combination of both leased and p.s.t.n. links, and very often the leased circuits are provided with the stand-by facility of using the p.s.t.n. when necessary (i.e. if the leased circuit should fail).

Two-wire and Four-wire Operation of Leased Data Links

The terms two-wire and four-wire refer to the line circuit that is presented to the data terminal. Fig. 9.1 shows the block diagram of a link that is operated two-wire over the whole of its length. Such an arrangement is only possible for a short line unless a low-bit speed ±80 V circuit is satisfactory. Higher speeds of transmission require the use of modems at each end of the line to convert the data into voice-frequency signals and Fig. 9.2 shows the circuit of an amplified data link. The line joining the two terminal repeater stations is operated on a four-wire basis for reasons discussed elsewhere [TSII] but, since the local lines connecting the data terminals to their nearest repeater stations are operated two-wire, the circuit as a whole is said to be two-wire presented. The four-wire section of the circuit is very likely to be routed over a channel in a multi-channel system.

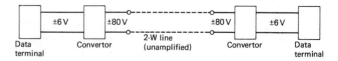

Fig. 9.1 2-wire d.c. data circuit

The two-wire sections of the circuit are only able to transmit 600 bits/sec or 1200 bits/sec signals in one direction at a time (half-duplex) since the frequencies used are 1300 Hz and either 1700 Hz or 2100 Hz and these occupy most of the available frequency band. Full-duplex operation is only possible if the return direction of transmission is operated at a different frequency and at a lower bit rate. Alternatively, full-duplex operation is possible if both directions of transmis-

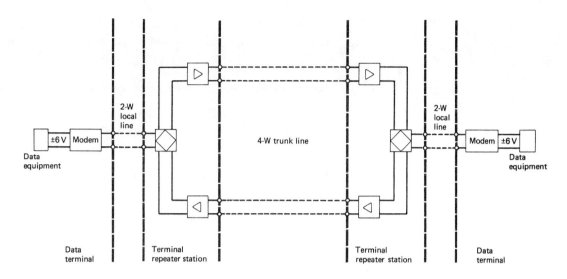

Fig. 9.2 2-wire data circuit using modems and amplifiers

Table 9.1

| | Frequency (Hz) | |
Channel	Binary 1	Binary 0
1	980	1180
2	1650	1850

sion are operated at the lower speed of 300 bits/sec. For the 300 bits/sec duplex operation the C.C.I.T.T. recommended frequencies are given in Table 9.1. The two directions of transmission are operated in different frequency bands and so they can be sorted out at the receiver by means of suitable filters.

When a data link is operated on a four-wire basis, two pairs of conductors are extended from the local repeater station right up to the modem in the data terminal as shown by Fig. 9.3. The link is now completely stable and provides two separate channels between the two data terminals, each of

Fig. 9.3 4-wire data circuit

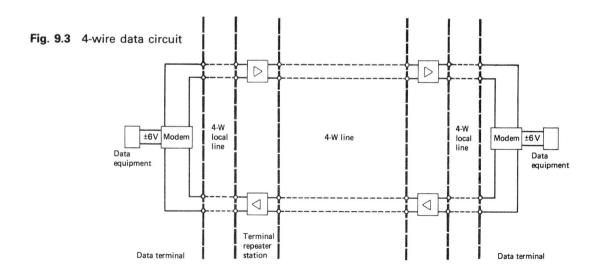

which is able to carry high-speed data simultaneously. Thus, the four-wire link provides full duplex high-speed operation.

Multiplexers, Front-end Processors, and Concentrators

Because of the high cost of line plant it is desirable to be able to use a line to carry more than one data link. *Multiplexers* and *concentrators* are two equipments which, in their different ways, increase the utilization of a point-to-point circuit. Further savings in costs are obtained with a concentrator since fewer modems and interfacing equipments are then required. A front-end processor relieves the digital computer of many tasks and allows time-sharing of the computer.

Multiplexers

A **multiplexer** combines several individual data links together using either *frequency division multiplex* (f.d.m.) or *time division multiplex* (t.d.m.). With **frequency division multiplex** [TSII], each data channel is shifted, or frequency translated, to a different part of the available frequency spectrum. The particular frequency to which a channel is shifted is determined by the frequency of the carrier wave which is modulated by that channel. For example, a 300–3400 Hz bandwidth point-to-point circuit can accommodate 12 110 bits/sec data channels. The bandwidth needed per channel is 55 Hz but, because of the need for interchannel frequency gaps to allow the filter to build up its attenuation, the carrier frequencies are spaced 240 Hz apart.

With **time division multiplex** the data channels each occupy the same frequency band but they are applied in time sequence to the line. Fig. 9.4 illustrates the basic principle of a time division multiplexer. Four data channels are shown operating at 300 bits/sec. The duration of a bit is $\frac{1}{300}$ seconds or 3.33 ms and so a 7-bit character occupies a time slot of 23.31 ms. Suppose that the common line is to be operated at the higher speed of 1200 bits/sec so that each bit sent to line will have a time duration of 0.83 ms. The data present on the channel 1, 2, 3 and 4 inputs to the t.d.m. system are fed into the appropriate channel buffer stores and held there until the store is given access to the common line. The clock pulses 1, 2, 3 and 4 are applied to each gate in turn to sample the stored information held in the associated buffer store and apply it to the common line for a time period equal to the duration of a character, i.e. 23.31 ms. When, for example, clock pulse 1 enables store A for 23.31 ms, the stored data is transmitted to line at the 1200 bits/sec bit rate. Clock pulse 1 then ends and

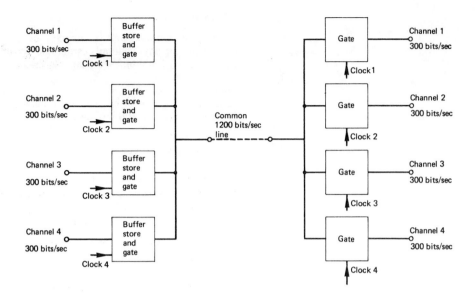

Fig. 9.4 Use of time division multiplexers

buffer store A is taken off line. Clock pulse 2 now enables
store B to allow one character of the data stored to be sent to
line. After 23.31 ms buffer store B is inhibited and buffer store
C enabled and so on until all four channels have been connected, in turn, to the common line. Clock pulse 1 now
enables buffer store A again and another character of the
stored data can be transmitted to the line and so on.

Clearly, synchronization between the transmitting and receiving equipments is essential in order that the clock pulses at
the receiver occur at the correct intervals in time. Otherwise
the pulses proper to one channel may well be routed to
another channel.

Usually, the output of a t.d.m. equipment will be applied to a
modem to convert the digital signals into voice-frequency
signals for transmission over the telephone network (see Fig.
9.5). With a t.d.m. system the bit rate on the common highway
is the *sum* of the individual channel bit rates.

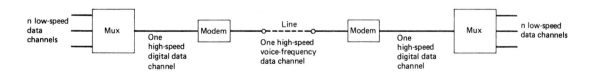

Fig. 9.5 Multiplexed voice-frequency data circuit

EXAMPLE 9.1

A 2400 bits/sec data link is to carry one 1200 bits/sec and a number of 300 bits/sec data channels using time division multiplex. Determine the number of 300 bits/sec channels that can be transmitted.

Solution
Line bit rate = sum of channel bit rates = 2400 bits/sec.
Hence, number of 300 bits/sec channels = 1200/300 = 4.

Front-end Processors

A digital computer generally interfaces with a data communication system through a **front-end processor** (f.e.p.). The f.e.p. relieves the computer of some tasks it would otherwise have to perform and enables its computing power to be devoted to the processing and storage of data. The f.e.p. performs the following tasks:

(1) The f.e.p. act as a *multiplexer* to allow several data channels to have access to the computer on a time-sharing basis. One or more high-speed channels connect the f.e.p. to the computer, while the other side of the f.e.p. is connected to a greater number of data links.

(2) The f.e.p. acts as a communications controller. It controls all the telecommunications facilities of the computer, monitoring all the associated modems to determine when any of them has data ready for processing. The f.e.p. decides when a modem shall have access to the computer itself and so ensures that the computer is not overloaded by a large number of messages arriving at the same time. The f.e.p. sets up all the necessary connections via the p.s.t.n. and meters all incoming calls and then produces bills for the use made of the computing facilities.

(3) The f.e.p. **polls** or asks each data terminal in turn whether it has any data to transmit to the computer. The interrogated terminal can reply by sending its data or it can signal that it has none to send. The f.e.p. can then check whether it has any data to send to that data terminal and if so transmits it. The f.e.p. will then poll the next terminal in the laid-down sequence. The polling can be carried out at such a speed that the f.e.p. acts as a multiplexer to provide time-sharing of the computer.

The Concentrator

The operation of a **concentrator** relies upon the fact that data is not normally transmitted continuously over a data link but, instead, is sent in *bursts* of varying lengths. A line concentrator is able to connect any of x inputs to any of y outputs, where $x > y$. The idea is illustrated by Fig. 9.6 in which any input

Fig. 9.6 Use of a line concentrator

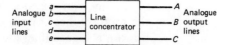

line *a*, *b*, *c*, *d* or *e* can be connected to any output line *A*, *B* or *C*. The concentrator polls each of its inputs in turn to discover whether it has data waiting to be transmitted. If it has, that input is switched through to an unoccupied output line and the distant modem is signalled that data transmission is about to commence. When incoming data is received via one of the lines *A*, *B* or *C*, the concentrator determines the destination data terminal and addresses the data.

Data concentrators include storage facilities which allow the input data to be stored and then read out from the store and transmitted down the line in blocks when the concentrator is polled by the f.e.p. at the computer centre. The storage facility allows data to be sent in smooth blocks or trains of data (which may well be the combination of the data originating from more than one terminal) and allows the use of a main line whose bit rate is less than the sum of the input bit rates. This does mean, of course, that if all the input lines are sending data at the same time, some of the data will have to be stored and there will be some delay in the concentrator before all of the data is passed on to the main line. Fig. 9.7 shows how a data concentrator is used; it should be noted that the concentrator works with digital signals only and any analogue signals must be changed into ±6 V signals before entering the circuit.

Sometimes a line concentrator is used before a multiplexer to achieve the the maximum utilization of the transmission path and Fig. 9.8 gives an example of this technique. For each

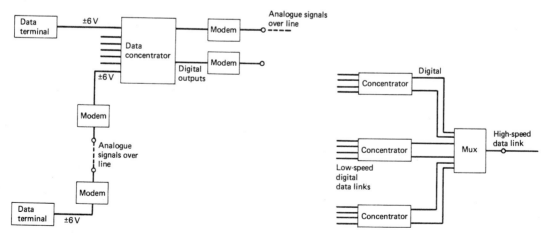

Fig. 9.7 Use of a data concentrator

Fig. 9.8 Use of concentrators and multiplexers

of the line concentrators the bit rate at its output is less than
the total bit rate at its input but for the multiplexer the output
bit rate is higher than the input bit rate on each input.

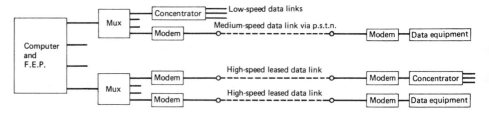

Fig. 9.9 A data network including both multiplexers and data
concentrators

A multiplexer can also be used as a simple kind of front-end
processor to permit time-sharing of a digital computer and Fig.
9.9 shows a possible data network. The f.e.p. polls each of the
concentrators sequentially and during the short time a concentrator is connected to the computer it can pass data into, or
receive data from, the computer. Clearly, the concentrator
must possess some storage facility so that it can store the data
received when the computer is dealing with another concentrator, or a direct link. As shown in the figure some concentra-
tion may be provided at the computer centre but other concentrators may be located considerable distances away.

Noise and Interference

All communication systems, be they line or radio, will always
contain unwanted noise voltages in addition to the wanted
signal voltages. The unwanted voltages are collectively known
as *noise* and may originate from a variety of different sources.
Most of these noise sources have been discussed elsewhere
[EIII] and will only be listed in this chapter.

(1) **Thermal agitation or resistance noise** is the inherent
noise voltage generated within any conductor because of random electron movement. The term conductor includes the
resistance of each conductor of a transmission line, resistors, the base-spreading resistance of a bipolar transistor, and
so on.
(2) **Semiconductor noise** is noise voltages generated within
any semiconductor for various reasons, which result in random
fluctuations in the flow of charge carriers (holes and/or elec-
trons) in the semiconductor material.
(3) Noise voltages caused by **faulty joints** in transmission
lines, in distribution frames, and in electronic equipment. A
faulty joint can lead to abrupt changes in the current flowing

across the joint and such a change can produce new and unwanted frequencies which can cause interference to the wanted signal.

(4) **Unwanted couplings**, magnetic and electric, with nearby power cables and electric railways may cause power frequency voltages to be induced into a cable pair.

(5) **Crosstalk** between conductors within a telephone cable can occur because of capacitance unbalances between the pairs.

(6) **Intermodulation** noise can be produced in multi-channel systems.

(7) **Impulse noise voltages** can be introduced into a data circuit routed over the p.s.t.n. because of couplings between circuits via the common impedance of the telephone exchange battery.

(8) A connection is set up in the telephone exchange by establishing electrical contact between switching points. **Mechanical vibrations** in the exchange equipment are likely to cause the efficiency of these contacts to vary and this is another way in which noise voltages or impulses can be generated.

(9) **Short breaks** in the transmission path are always likely to occur for a number of reasons; these breaks are normally only of some milliseconds duration and are not noticeable in a telephone speech connection. For a data link however, every short break in the transmission path which takes place means that one or more bits have been lost.

(10) **Sudden level changes** in amplified circuits, particularly those routed over a multi-channel telephony system, may cause sudden changes in their overall loss.

(11) **Frequency changes** The correct operation of a multi-channel system relies upon a carrier being re-inserted at the receiver at the correct frequency. Although elaborate synchronization circuitry is used, the frequency *off-set* may be as much as $\pm 2\,\text{Hz}$. When a data circuit is routed over several multi-channel systems in cascade, the overall frequency off-set could become much greater than $2\,\text{Hz}$; this effect leads to distortion of the received signal.

Leased circuits are not routed via telephone exchange switching equipment and so are not subject to the noise originating from exchange power supplies and switching contacts. Because of this a p.s.t.n. link is inherently noisier than a privately leased point-to-point circuit.

Continuous noise, such as resistance and semiconductor noise, is of fairly low level and it does not cause much, if any, increase in the error rate of the received data. *Impulse noise* is the general term for any noise sources which are not continu-

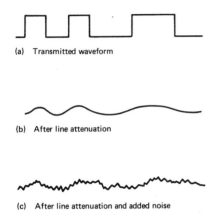

(a) Transmitted waveform

(b) After line attenuation

(c) After line attenuation and added noise

Fig. 9.10 Effect of noise on a d.c. data circuit

ous but instead produce sudden pulses of noise voltage in a circuit. Impulse noise can result from faulty joints, crosstalk in cables and exchange power supplies, particularly if dial impulses are induced, and telephone exchange switch contacts. The level of impulse noise can be large in comparison with the wanted signal level and is most troublesome when the line attenuation is large and/or the bit duration is short compared with the duration of the noise impulse.

The effect of superimposed noise voltages on a d.c. data signal is shown by Fig. 9.10. The rectangular-shaped transmitted data waveform is reduced in amplitude and rounded off by the effects of the line attenuation and filtering. The effect of the noise voltages is to make it more difficult for the receive equipment to respond to the incoming data with a low probability of error.

The use of frequency shift modulation or differential phase modulation increases the signal-to-noise ratio on a line because of the inherent properties of these forms of modulation [RSIII]. This means that the error probability is reduced.

Error Control

A data link is subjected to a number of sources of noise, mentioned in the previous section of this chapter, which can produce errors in the received data. More serious are the errors produced by brief breaks in the transmission path; the loss of a single bit may well cause an error which could alter the whole meaning of a message. The possible causes of interruptions are many and include the following: faulty or loose test connections, poorly soldered joints, changes in the power supplies from the main plant to the standby plant, mechanical vibrations, and, last but by no means least, breaks caused by technicians working on nearby circuits or equipments.

To reduce the error rate to an acceptable level (it is not economically possible to eliminate it), a data system will usually include some kind of error-checking mechanism and in some cases error correction as well. A number of error detection methods are in use in different networks but this book will only outline the principles of the most commonly employed method, known as the **parity system**.

Character parity means that each character signalled to the line has one extra bit added. The added bit may be either a 1 or a 0; the choice being made so that the *total* number of 1 bits transmitted per character is ODD if *odd parity* is used or EVEN if *even parity* is employed. The parity bit is transmitted at the end of each character. Two examples of this system are given below.

(i) Character 0 0 1 1 1 0 1; there are four 1 bits in this character and so the added parity bit must be a 1 if odd parity is used and a 0 if even parity is employed.

(ii) Character 1 0 1 0 1 1 1; there are five 1 bits in this case and so odd parity requires the addition of a 0 parity bit, while if even parity is used the added parity bit must be a 1.

At the receiver, the incoming character is checked to determine how many 1 bits it contains, and if the total is odd (or even) the received character is taken as being correct. Should the total number of 1 bits received be an even number when odd parity is used, or an odd number when even parity is used, an error in the received message is detected which must, in some way, be indicated at the receiving terminal or signalled back to the transmitter so that the data can be sent again.

Parity bit error checking is successful in locating any single-bit errors that should occur although clearly two bits in error would not be detected. However the likelihood of two bits in error in one character is acceptably small.

Two-bit errors can also be detected if the parity bit principle is extended to whole blocks of data. It is also possible to use parity checks other than the number of 1 bits received per character but such procedures are beyond the scope of this book. In practice, block parity systems are normally used and character parity is fairly rare.

With an *error correction* system it is usual to acknowledge each character or each block as it is received. Character-by-character acknowledgement is very time consuming when the error rate is low but, on the other hand, block error correction means repeating a complete block of data whenever an error does occur. For either error correction method, the next character of the next block is not sent by the transmitter until an Ack. signal is received from the distant receiver.

Typical Data Networks

A data network will contain one or more digital computers and a number of data terminals, each of which will have some kind of access to a computer. The operation of any data link may be either *on-line* or *off-line*. The basic idea of an **off-line data system** is shown by Fig. 9.11. The data to be processed is punched onto a paper tape (or paper card or magnetic tape) and the tape is fed into a tape reader. The tape reader

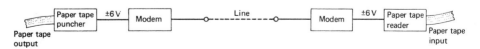

Fig. 9.11 Off-line data circuit

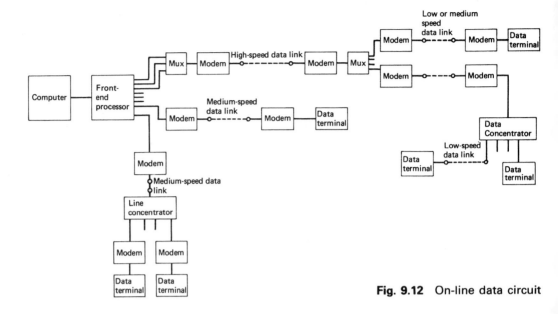

Fig. 9.12 On-line data circuit

converts the input data into ±6 V data waveform and this, in turn, is converted into voice-frequency signals by the modem. At the far end of the line the incoming signals are changed back into ±6 V signals which operate the paper tape puncher to punch holes in a paper tape to store the received data. The punched tape can then be stored until its processing is to be carried out, when it is fed into a tape reader having direct access to a computer.

An **on-line data terminal** has instant, and apparently, exclusive access to a computer and the layout of a possible system is shown by Fig. 9.12. The network shown is a composite of the various techniques outlined earlier in this chapter.

The reasons for the existence of computer bureaux have previously been mentioned and Fig. 9.13 shows how a bureau network might be set up. A number of data terminals are able to operate simultaneously (apparently) with the remote computer using the p.s.t.n. or, in some cases, a privately-leased circuit. The network makes computing power available, on request, to a large number of customers located anywhere in the network.

When a link is to be established over the p.s.t.n. the caller presses the TELE button on his telephone receiver and then dials the number of the computer bureaux. Once the call has been answered, the DATA button on the telephone is pressed and control of the connection passes to the modem. The data number of the caller is then transmitted to *log-in* to the computer. Once the computer has signalled that the logging-in is satisfactory, the transmission of data can begin.

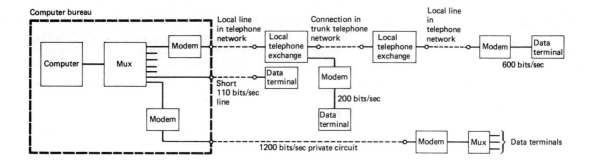

Fig. 9.13 Computer bureau network

The traffic pattern of the calls made to a computer bureau is quite different to that experienced with ordinary telephone calls. The average telephone call lasts for three minutes, the average call to the computer bureau lasts for about 25 minutes, but many data calls are maintained for an hour or even longer. This means that the traffic flow on the local lines and the telephone exchange equipment is very much heavier than for the same number of lines and equipment serving telephone customers. The increased traffic loading means that a larger number of junction lines are needed between the local exchange and other exchanges than would be provided for a telephone exchange of the same size serving exclusively telephone customers. Also, a larger number of final selectors, and to some extent other switches too, are needed to serve the lines to the bureau—the number of final selectors found necessary may even exceed the number of lines provided to the bureau.

The long average duration of a data connection is not important to a user when only a local telephone call is needed to gain access to the computer bureau but it could prove to be expensive when long-distance trunk calls are involved. To overcome this problem a multiplex (frequency-division or time-division) system can be provided on a leased private circuit from the computer bureau to a convenient remote (from the bureau) point. The customer at the remote point can then make a local call to gain access to his end of the multi-channel system and thence to the bureau.

One of the more common kinds of data network is the kind where data originating from a large number of variously located branch offices is to be processed by a central computer, stored, and when needed is transmitted back to the originating or some other branch office. This is, of course, the type of network used by the high-street banks. Each branch office has access to a central computer which keeps details of customer's

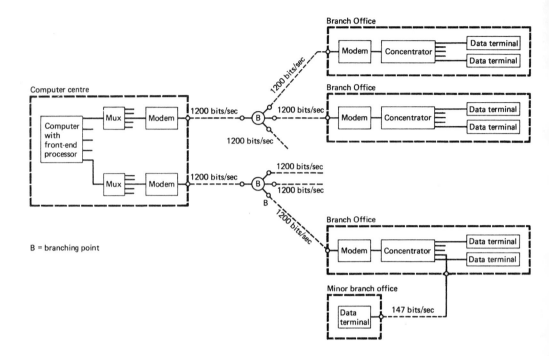

Fig. 9.14 Centralized accounting data network

accounts, standing orders, direct debits, and so on, and prints out bank statements, pays standing orders and direct debits by transferring data from one location to another. When a payment is made from an account with one bank to an account held with another bank, a computer-to-computer transfer of data will be needed.

Fig. 9.14 shows the layout of a centralized accounting data system. The demands of this type of network are normally satisfied by the use of 1200 bits/sec data links with some 147 bits/sec links between minor branches and larger branches. The branching points allow a number of data terminals to share a common line without any change in the bit rate, the terminals are polled by the f.e.p. and are signalled when they can transmit their data. Should a branch be unable to establish contact with the computer over its direct route for some reason, a link can sometimes be set up instead over the p.s.t.n., although it may then be necessary to transmit at the reduced speed of 600 bits/sec. In the United Kingdom this facility is not permitted if the modems used are not Post Office types.

When two computers are to communicate directly with one another they can be directly linked together without interface equipment only when the distance between them is very short.

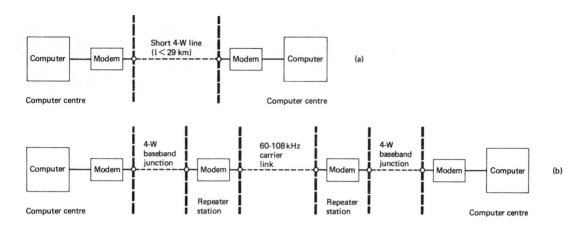

Fig. 9.15 Inter-computer links: (*a*) short distance, (*b*) longer distances

For longer distances the arrangements shown in Fig. 9.15 is employed. The modem encodes the data outputed by the computer into a random pattern. This is because repetitive data patterns will have large-amplitude components in their frequency spectra and these may cause interference with other pairs in the cable. The random data pattern generated by the modem is band-limited to 1.92–36 kHz (known as the *baseband*). The maximum distance that can be linked in this way is limited, by the line attenuation at 36 kHz, to 29 km. For longer distances than this, the baseband signal is transmitted to the nearest repeater station and there it is used to vestigial-sideband amplitude-modulate a 100 kHz carrier wave. The modulation process shifts the baseband signal to the C.C.I.T.T. carrier telephony baseband of 60–108 kHz and so the v.s.b. a.m. signal can readily be sent over a carrier link (see Fig. 9.15*b*).

Computers are also widely used by other organizations of many kinds, such as gas, water, and electricity authorities, and, of course, the police. In the United Kingdom, for example, a central computer is used to store police records of criminals and of stolen property such as motor cars, and on missing persons. Clearly it is convenient for the police if every police-man, on foot, in a car or in an office, can have immediate access to the computer to obtain any necessary information. The basic way in which this is achieved is shown by Fig. 9.16. All the police headquarters in the country have access to the centralized police computer and all policemen have access to a data terminal. This access may be via a radio link or a telephone call as shown in the figure or, when it is convenient, by personal contact with the data operators.

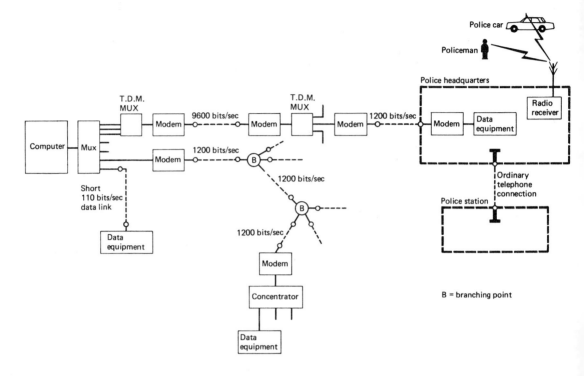

Fig. 9.16   Police computer network

Exercises

9.1. Explain why there is a need for synchronization in a data link. Draw the data waveform, including start and stop signals, when an 8-bit character is transmitted. If the stop signal is twice the unit length and the complete character is sent in $\frac{1}{10}$ sec, calculate (i) the bit rate, (ii) the baud speed.

9.2. What is the difference between a concentrator and a multiplexer? Draw the block diagram of a data system in which two leased circuits and two links routed via the p.s.t.n. are connected to a multiplexer for transmission over a high-speed data link to another multiplexer which is connected to the front-end processor of a computer. Label each data link drawn with a typical bit rate.

9.3. List the factors which determine whether a leased point-to-point circuit or the p.s.t.n. should be used to connect a data terminal to a computer centre. Discuss how the decision will be affected by the use of (i) multiplexers, (ii) concentrators.

9.4. Explain the need for point-to-point data links (i) between a computer and a data terminal and (ii) between two computers.

Show, with the aid of block diagrams, how two computers can be linked together when the distance between them is (i) very short, (ii) a few kilometres, (iii) 40 km.

9.5. Draw the block schematic diagram of a centralized computer network. Include examples of the use of both multiplexers and concentrators in your diagram. Why are branching points often used?

9.6. What is a computer bureau? Explain the effects on the traffic flow through a telephone exchange that has a number of lines to a computer bureau.

9.7. Draw sketches to explain what is meant by the terms two-wire operation and four-wire operation when applied to (i) amplified audio-frequency lines (ii) data links. Give some reasons why a four-wire circuit is often used for data transmission.

9.8. List the sources of noise in a data transmission system. Say which of these sources are *impulse* noise. Explain how impulse noise limits the maximum possible speed of transmission.

9.9. Explain the need for point-to-point links in a data network. What are the advantages of leasing a private circuit? Give some possible ways in which a point-to-point link may be routed. Explain why modems and amplifiers are used for most long distance data links.

9.10. Draw block diagrams to show the difference between two-wire and four-wire presented data circuits and briefly explain their operation. What are the advantages of four-wire operation?

Short Exercises

9.11. Twelve 110 bits/sec data channels are simultaneously transmitted over a line using time division multiplex. What is the bit rate on the line?

9.12. Six 110 bits/sec data links are connected to a concentrator which has three output links. What is the bit rate on the output links?

9.13. Explain briefly why 600 bits/sec duplex operation cannot be obtained using two-wire operation of a data link.

9.14. State the advantages of voice-frequency operation of a data link compared with d.c. signalling over long distances.

9.15. How many cycles of 1300 Hz signal are there in the time for one bit in a 1200 bits/sec system?

9.16. Discuss briefly the circumstances which may decide that a data link should be rented full-time and not dialled-up as required.

9.17. The bit trains (i) 1 1 0 1 1 1 1 and (ii) 0 0 1 1 0 1 0 are to be transmitted. Add a parity bit to each train assuming (*a*) odd parity and (*b*) even parity.

9.18. Explain briefly the effects on the traffic flow through a telephone exchange of a computer bureau.

9.19. Draw a block diagram to show how two computers can be connected directly together (i) when the distance between them is very short, (ii) when the distance between them is many kilometres.

9.20. Why is impulse noise more of a problem in a data network than continuous noise? Give some sources of impulse noise.

10 Pulse Code Modulation

A continuing demand over a number of years for more and more junction and trunk circuits has led to the widespread use of multiplex telephone systems using either frequency division or time division multiplex. For shorter-distance junction and trunk circuits, a considerable increase in the traffic-carrying capacity of a pair in an audio-frequency cable can be obtained if time division multiplex using *pulse code modulation* (p.c.m.) is used. The standard systems used in the United Kingdom provide 24 channels (the earlier system) or 30 channels over (usually) 0.63 mm cable pairs, from which all the loading coils have been removed. Higher-capacity systems, e.g. 1680 channels, are also available and are presently being introduced into the U.K. trunk telephone network.

Pulse code modulation can also be applied to radio links but the development of such systems is not so far advanced.

Pulse Amplitude Modulation

A time division multiplex system is based upon the sampling of the amplitude of the information signal at regular intervals, and the subsequent transmission of one or more pulses to represent each sample. In an *analogue* pulse system, for the intelligence contained in the information signal to be transmitted the characteristics of the pulse must, in some way, be varied in accordance with the amplitude of the sample. The pulse characteristic which is varied can be the amplitude, or the width, or the position of the pulse, to give either pulse amplitude, pulse duration, or pulse position modulation. For a *digital* pulse system, i.e. p.c.m., information about each sample is transmitted to the line by a train of pulses which indicate, using the binary code, the amplitude of the sample. The analogue methods of pulse modulation are rarely used in their own right in modern systems but pulse amplitude modulation (p.a.m.) is employed as a step in the production of a p.c.m. signal.

With p.a.m., pulses of equal width and spacing have their amplitudes varied in accordance with the characteristics of a modulating signal. Fig. 10.1a shows an unmodulated pulse train, often known as the *clock*, which has a periodic time of T seconds. Thus the number of pulses occurring per second—known as the pulse repetition frequency—is equal to $1/T$. The clock and the modulating signal are applied to the two inputs

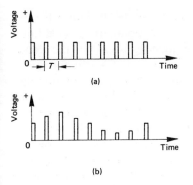

Fig. 10.1 Pulse amplitude modulation: (a) unmodulated pulses, (b) p.a.m. wave

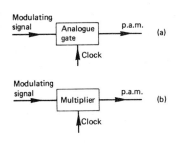

Fig. 10.2 Production of a p.a.m. signal: (a) use of an analogue gate, (b) use of a multiplier

of an analogue gate, or a multiplier. The gate, or the multiplier, produces an output signal, equal to the instantaneous value of the modulating signal, only when a clock pulse is present (see Fig. 10.2). Thus the p.a.m. output of the analogue gate (or the multiplier) consists of successive *samples* of the modulating signal and, assuming a sinusoidal modulating signal, is shown in Fig. 10.1b.

The p.a.m. waveform contains components at a number of different frequencies:

(i) The modulating signal frequency.
(ii) The pulse repetition frequency and upper and lower sidefrequencies centred about the pulse repetition frequency.
(iii) Harmonics of the pulse repetition frequency and upper and lower sidefrequencies centred upon each of these harmonics.
(iv) A d.c. component whose voltage is equal to the mean value of the p.a.m. waveform.

The spectrum diagram of a p.a.m. waveform is shown in Fig. 10.3a. The modulating signal and each of the sidebands are represented by truncated triangles in which the vertical ordinates are made proportional to the modulating frequency, and no account is taken of amplitude. This method of representing signals shows immediately which sideband is erect and which is inverted [TSII]. The modulating signal occupies the frequency band $f_2 - f_1$ and so the upper and lower sidebands of the pulse repetition frequency are

$$(f_s + f_2) - (f_s + f_1) \quad \text{and} \quad (f_s - f_1) - (f_s - f_2)$$

The frequency spectrum of a p.a.m. waveform contains the *original modulating signal* and this means that demodulation can be achieved by passing the p.a.m. waveform through a

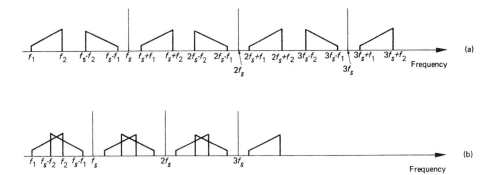

Fig. 10.3 Spectrum diagrams of a p.a.m. signal: (a) sampling frequency f_s greater than twice the highest modulating frequency (b) sampling frequency f_s less than twice the highest modulating frequency

Fig. 10.4 Demodulation of a p.a.m. waveform

low-pass filter (see Fig. 10.4). The low-pass filter must be able to pass the highest frequency f_2 in the modulating signal but it must *not* pass the lowest frequency $f_s - f_2$ in the lower sideband of the pulse repetition frequency. Clearly, for this to be possible the lower sideband must not overlap the modulating signal in the way shown by Fig. 10.3b. This means that the pulse repetition frequency, also known as the **sampling frequency** f_s, must be *at least twice* the highest frequency contained in the modulating signal; this requirement is known as the **sampling theorem**. In practice, the sampling frequency must be somewhat higher than twice the highest modulating signal frequency in order to provide a frequency gap in which the filter can build up its attenuation. For example, the Post Office 30-channel p.c.m. system uses a 8000 Hz sampling frequency for a highest modulating signal frequency of 3400 Hz.

Pulse amplitude modulation is rarely used in its own right because unwanted noise and interference voltages vary the amplitudes of the pulses and degrade the system signal-to-noise ratio. The system is, however, an essential part of the process of producing a pulse code modulated signal.

Pulse Code Modulation

In pulse code modulation the total audio-frequency amplitude range to be transmitted by the system is divided into a number of allowable voltage levels. Each of these levels is allocated a number as shown by Fig. 10.5 in which 8 levels have been drawn. The analogue signal is sampled at regular intervals to produce a pulse amplitude modulated waveform, but then the pulse amplitudes are rounded off to the nearest allowed voltage level. The process of approximating the sampled signal amplitudes is known as **quantization** and the allowed voltage levels are called quantization levels. The number of the nearest quantization level to each sampled amplitude is encoded into the equivalent binary form and it is then transmitted to the line. Usually, a pulse is sent to represent binary 1 and no pulse is sent to represent binary 0.

Consider Fig. 10.5. The signal waveform is sampled at time intervals t_1, t_2, t_3, etc. At time t_1 the instantaneous signal amplitude is between levels 5 and 6 but, since it is nearer to level 6, it is approximated to this level. At instant t_2 the signal voltage is slightly greater than level 6 but is again rounded off to that level. Similarly, the sample taken at time t_3 is represented by level 2, the t_1 sample by level 1, and so on. Each quantized sample is then *encoded* into the binary pulses shown alongside. The binary pulse train which would represent this signal is shown in Fig. 10.6. A space, equal in time duration to one bit, has been left in between each binary number in which synchronization information can be transmitted.

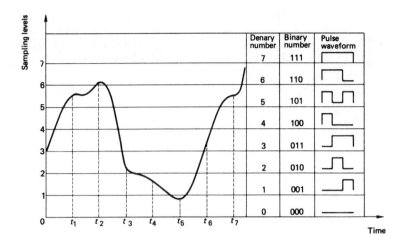

Fig. 10.5 Quantization of a signal

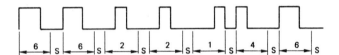

Fig. 10.6 Binary pulse train representing the signal shown in Fig. 10.5

Since the signal information is transmitted in binary form, the number of levels used is always some power of 2^n starting from zero; thus the highest numbered level is $2^n - 1$, where n is the number of bits used to represent each sample. In Fig. 10.5, eight or 2^3 levels only were drawn for clarity but practical systems employ many more quantization levels. For example, the Post Office 30-channel system has 256 levels so that $n = 8$, whilst the 24 channel system uses 128 levels with $n = 7$.

Quantization Noise

The signal received at the far end of the system is not an exact replica of the transmitted signal but, instead, is the quantized approximation to it. Fig. 10.7 shows a particular modulating signal and its quantized approximation. The difference between the two is an error waveform which, since it sounds noisy at the output of the system, is known as **quantization noise**. The magnitude of the quantization error depends upon the number of quantization levels used *and* the sampling

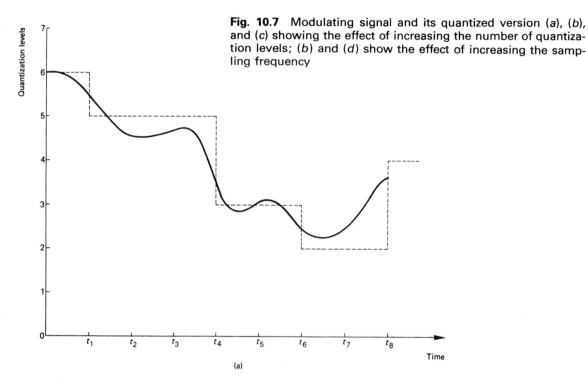

Fig. 10.7 Modulating signal and its quantized version (a), (b), and (c) showing the effect of increasing the number of quantization levels; (b) and (d) show the effect of increasing the sampling frequency

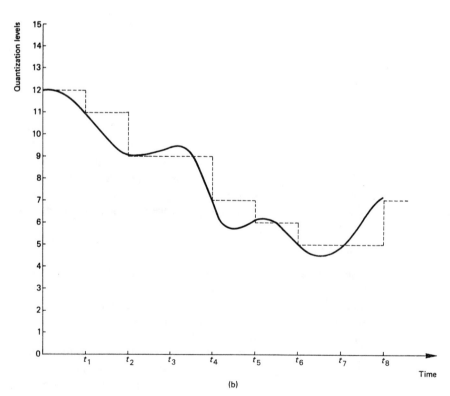

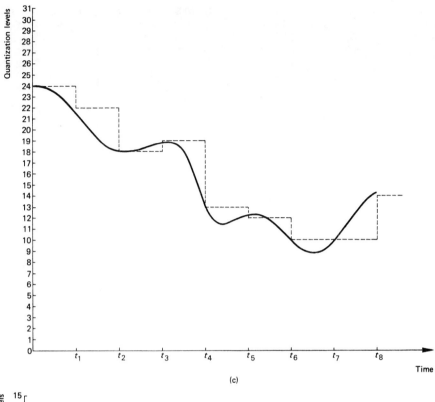

(c)

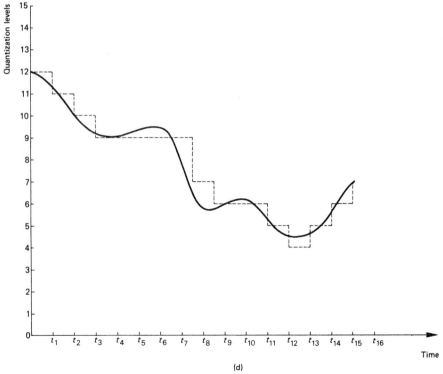

(d)

Fig. 10.8 Error waveforms for Fig. 10.7

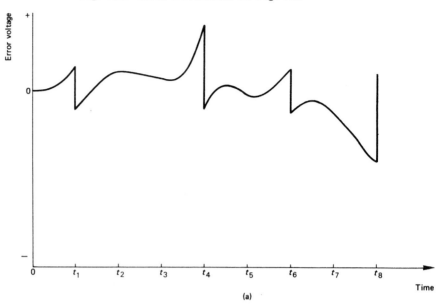

(a)

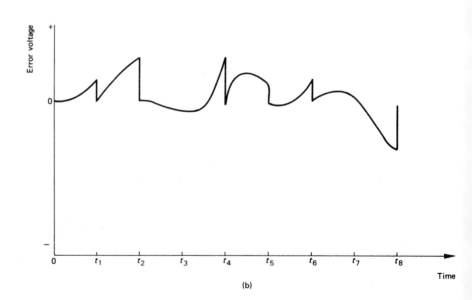

(b)

frequency; increase in either or both of these parameters produces a reduction in the error. This is illustrated by Fig. 10.7 in which the effect of increasing the number of quantization levels from 8 to 16 and then to 32 is shown by *a*, *b* and *c*. Also Figs. 10.7*b* and *d* show the effect of an increase in the

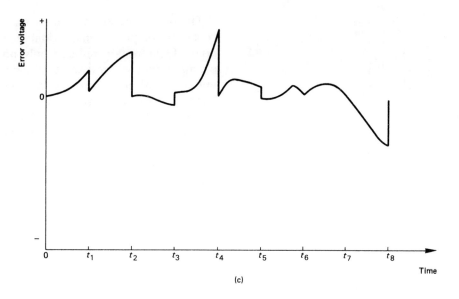

(c)

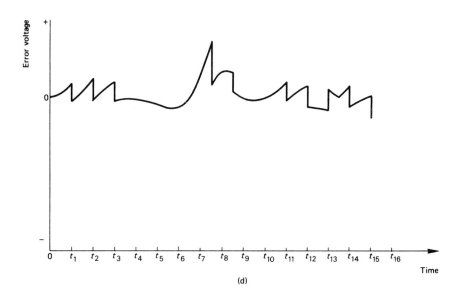

(d)

sampling frequency. The error waveforms for these four cases
are given in Fig. 10.8.

Unfortunately, as will be shown later, increasing the number
of quantization levels and/or increasing the sampling frequency results in a wider bandwidth requirement.

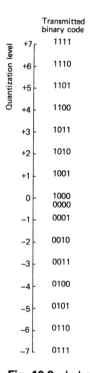

Quantization level

	Transmitted binary code
+7	1111
+6	1110
+5	1101
+4	1100
+3	1011
+2	1010
+1	1001
0	1000
	0000
−1	0001
−2	0010
−3	0011
−4	0100
−5	0101
−6	0110
−7	0111

Fig. 10.9 Labelling the quantization levels to indicate positive or negative voltage
Left-hand digit = 1   positive voltage
Left-hand digit = 0   negative voltage

The output waveform of a p.c.m. system can be regarded as consisting of the original modulating signal plus quantization noise. Quantization noise is only present at the output of a p.c.m. system when a signal is being transmitted. The maximum quantization error is the same for all amplitudes of signal and so the signal-to-noise ratio at the output is worse for the smaller amplitude signals. Some improvement can be realized if the first bit is used to indicate whether the sampled voltage is positive or negative with respect to earth. Thus, if four bits are used the quantization levels could be numbered in the manner shown in Fig. 10.9.

Non-linear Quantization

The use of equally-spaced quantization levels as so far assumed is alright as long as the signal amplitude is large enough to cover several quantization levels. For small-amplitude speech signals, however, the output signal-to-noise ratio may well be inadequate. The signal-to-noise ratio for small signals can be improved, without increasing the number of quantization levels, by the use of non-linear quantization. The quantization levels are no longer equally spaced but, instead, are much closer together near the middle of the amplitude range than near the two ends of the range. This will ensure that the small-amplitude signals are precisely quantized, while the larger-amplitude signals are quantized more coarsely. The spacing of the quantization steps is arranged to follow a logarithmic law such that a more or less constant output signal-to-noise ratio is obtained for signals of all allowable amplitudes.

The improvement in the quantization accuracy that non-linear quantization can give is illustrated by Fig. 10.10. The same small-voltage signal has been applied to a linear quantizer in Fig. 10.10a and to a non-linear quantizer in Fig. 10.10b. It is clear that the quantization error is much smaller in the second case.

Non-linear quantization can be obtained in two different ways:

(1) the analogue signal can have its range of amplitudes compressed and then encoding can take place, or
(2) a non-linear encoder can be employed.

The Bandwidth of a Pulse Code Modulated Waveform

The allowable amplitude range of a pulse code modulation system is divided into 2^n quantization levels. Each time the analogue signal is sampled, n binary pulses are transmitted to

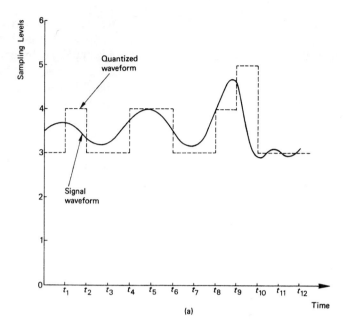

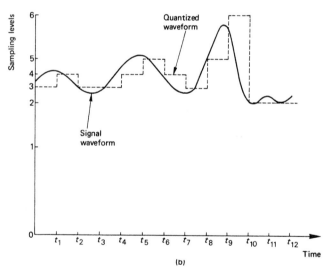

Fig. 10.10 Showing the improvement in quantization accuracy obtained by the use of non-linear quantization: (a) linear, (b) non-linear quantization

indicate the number of the appropriate quantization level. Suppose that each train of pulses representing one sample is accompanied by one synchronization pulse. Hence each sample is represented by $n+1$ binary pulses or bits. The analogue signal is sampled f_s times per second, where f_s is the sampling frequency, and so the number of bits transmitted per sample is $(n+1)f_s$.

$$\text{Bit rate} = (n+1)f_s \qquad (10.1)$$

The maximum fundamental frequency of the transmitted binary pulse train occurs when it consists of alternate 1s and 0s as shown in Fig. 10.11. The maximum fundamental frequency is the reciprocal of the periodic time of the p.c.m. waveform and is equal to one-half of the bit rate. The minimum frequency of the p.c.m. waveform occurs when the binary number consists of consecutive 1 or 0 bits and is equal to 0 Hz. The bandwidth occupied by the p.c.m. waveform, assuming that only the fundamental frequency component need be transmitted is thus from 0 Hz to (bit rate)/2 Hz.

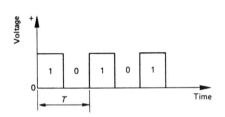

Fig. 10.11 P.C.M. waveform having the highest fundamental frequency

$$\text{Bandwidth} = \frac{\text{bit rate}}{2} \qquad (10.2)$$

EXAMPLE 10.1

Calculate the minimum bandwidth needed for a p.c.m. system which uses 128 quantization levels, assuming that each sample signalled to line is accompanied by one synchronization bit. The sampling frequency is 8 kHz.

Solution
128 quantization levels can be described by 7 bits and so the number of bits transmitted per sample is 8. Therefore

$$\text{Bit rate} = 8 \times 8000 = 64 \text{ kilobits/sec}$$
$$\text{Bandwidth} = 64 \times \tfrac{1}{2}10^3 = 32 \text{ kHz}$$

Time Division Multiplex

The general principles of time division multiplex have been described earlier in this book and elsewhere [TSII]. Here t.d.m. will be discussed in terms of the C.C.I.T.T. 30-channel p.c.m. system now standardized by the British Post Office.

In the 30-channel p.c.m. system, each sample of the information signal is represented by 8 bits, the first bit indicating the polarity of the sampled amplitude. Each bit is 488 ns wide and, hence, each sample occupies a time period of $8 \times 0.488 = 3.9\ \mu s$. A sample is taken every 1/8000 sec or every 125 μs, and this means that the larger part of each sample period is unoccupied. The unused time periods can be used to carry other p.c.m. channels; the number which can be fitted into the time available is $125/3.9 = 32$ channels. Two of these "channels" are used for synchronization purposes (that is channels 0 and 16) and so 30 channels are made available to carry speech signals.

The channels are interleaved sequentially, sample by sample, and the time period occupied by 32 time slots is known as a **frame**. Thus a frame occupies 125 μs and contains $32 \times 8 = 256$ bits. The line bit rate is

$$256 \times 8000 \quad \text{or} \quad 2048 \text{ kilobits/sec}$$

[The line bit rate of the 24-channel system is $24 \times 8 \times 8000 = 1536$ kilobits/sec.]

Line Signals

The information about the sampled amplitudes is signalled to the line using the binary code. So far it has been assumed that unipolar pulses are transmitted, i.e. a positive voltage representing binary 1 and zero voltage representing binary 0. This line code has the disadvantages that (i) it always has a d.c. component, (ii) it contains low-frequency components of relatively large amplitude. The d.c. and low-frequency components are not wanted because their absence simplifies the design of the pulse regenerators which are fitted along the length of the line.

The disadvantages of unipolar signalling can be overcome by the use of **alternate mark inversion** (a.m.i.), and the principle of this method of signalling is shown by Fig. 10.12. The basic unipolar signal (Fig. 10.12a) has its alternate bits inverted as

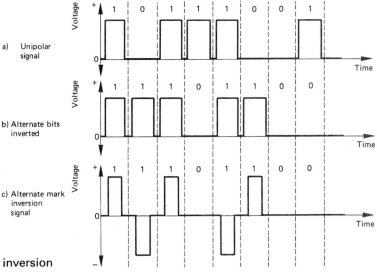

Fig. 10.12 Alternate mark inversion

shown by b and then (Fig. 10.12c) a bipolar signal is generated by inverting alternate *marks* or 1 bits. This procedure ensures that each 1 bit is of opposite sign to the preceding 1 bit. Alternate digit inversion reduces the probability of a long stream of 0s being transmitted to line when a channel is idle.

Alternate mark inversion signalling has the disadvantage that a long string of 0s in a bit stream results in no pulses being sent to line. The line pulse regenerators derive their timing information from the pulses incoming to them and so their operation is adversely affected by a period of no pulses. To overcome this problem a mark or 1 bit can be inserted into the bit stream after a given number of 0s have been sent. This kind of signalling is known as **high density bipolar n** (HDBn). In the 30-channel system, n is chosen to be 3, the encoding being known as HDB3. Thus 3 is the maximum number of consecutive 0s allowed in the transmitted signal.

Use of Pulse Regenerators

In its passage along a telephone line, the p.c.m. signal is both attenuated and distorted but, provided the receiving equipment is able to determine whether a pulse is present or absent at any particular instant, no errors are introduced. To keep the pulse waveform within the accuracy required pulse regenerators are fitted at intervals along the length of the line. The function of a **pulse regenerator** is to check the incoming pulse train at accurately timed intervals for the presence or absence of a pulse. Each time a pulse is detected, a new undistorted pulse is transmitted to line and, each time no pulse is detected, a pulse is not sent.

The simplified block diagram of a pulse regenerator is shown in Fig. 10.13. The incoming bit stream is first equalized and then amplified to reduce the effects of line attenuation and group-delay/frequency distortion. The amplified signal is applied to a timing circuit which generates the required timing pulses. These timing pulses are applied to one of the inputs of two two-input AND gates, the phase-split amplified signal being applied to the other input terminals of the two gates. Whenever a timing pulse *and* a peak, positive or negative, of the incoming signal waveform occur at the same time, an output pulse is produced by the appropriate pulse generator. It is arranged that an output pulse will not occur unless the peak signal voltage is greater than some pre-determined value in order to prevent false operation by noise peaks.

Provided the bit stream pulses are regenerated before the signal-to-noise ratio on the line has fallen to 21 dB, the effect of line noise on the error rate is extremely small. This means that impulse noise can be ignored and white noise (i.e. noise of

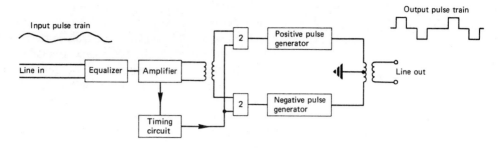

Fig. 10.13 Pulse regenerator

constant voltage over the operating bandwidth) is *not* cumulative along the length of the system. This feature is in marked contrast with an analogue system in which the signal-to-noise ratio must always progressively worsen towards the end of the system. Thus, the use of pulse regenerators allows very nearly distortion-free and noise-free transmission, regardless of the route taken by the circuit or its length.

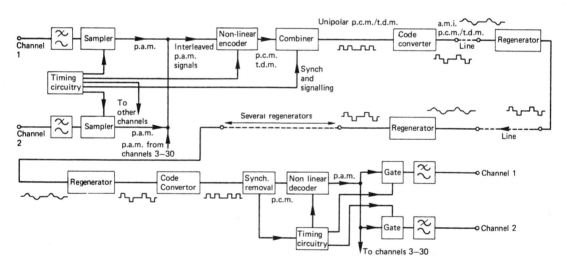

Fig. 10.14 30-channel p.c.m. system

Pulse Code Modulation Systems

The basic block diagram of a 30-channel p.c.m. system is shown by Fig. 10.14, in which 30 band-limited (to 4 kHz) channels are sampled sequentially at a sampling frequency of 8000 Hz to produce p.a.m. waveforms. The p.a.m. waveforms are interleaved on the common highway to produce a time division multiplex signal before the multiplex signal is encoded by the non-linear encoder to give a unipolar p.c.m. signal.

Synchronization and signalling bits are then added before the signal is converted to the HDB3 code and is then transmitted to the line at 2048 kbits/s. The transmitted bit stream is attenuated and distorted as it is propagated along the line and, to restore the pulse waveform, pulse regenerators are fitted at 1.828 km intervals along the length of the line.

Coaxial cables are well suited to the transmission of very high bit rate digital signals, and systems of 120 Mbits/sec are being introduced into the United Kingdom trunk network. Larger capacity p.c.m. systems will employ the C.C.I.T.T. 30-channel system as their basic building block.

The 24-channel p.c.m. system is very similar to the 30-channel system, the main differences being that, in 24-channel,

(i) A compressor followed by a linear encoder is used to get the required non-linear encoding.
(ii) The transmitted bit stream uses the alternate mark inversion code.
(iii) The line bit rate is 1536 kbits/sec.

Higher-order T.D.M. Systems

The C.C.I.T.T. 30-channel p.c.m. system can be used as the basic building block for higher-order t.d.m. systems in a similar manner to the way in which large-capacity f.d.m. systems are built up from the 12-channel group.

A 30-channel frame contains $32 \times 8 = 256$ bits and so the line bit rate is

$$256 \times 8000 = 2048 \text{ kbits/sec}$$

The higher-order bit rates recommended by the C.C.I.T.T. are

2nd order 8448 kbits/sec
3rd order 34 368 kbits/sec
4th order 139 264 kbits/sec

The 30-channel group can be represented by the block diagram of Fig. 10.15. The box marked *muldex* represents the multiplexing and demultiplexing equipment required for both directions of transmission.

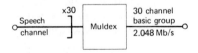

Fig. 10.15 Block diagram of a 30-channel p.c.m. system

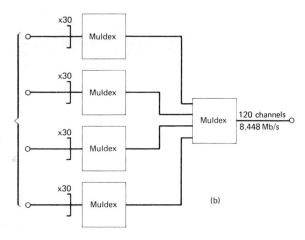

Fig. 10.16   Block diagram of a 120-channel p.c.m. system

Four 30-channel groups can be combined to form a 120-channel 8448 kbits/sec system (Fig. 10.16). Each of the four groups is fed into a store, the contents of which are fed into line in sequence. It should be noted that 4×2048 is 8192 kbits/sec. Some of the surplus bits are used for synchronization purposes and supervisory signals and the remainder are redundant.

The Transmission of Data Signals over P.C.M. Links

Data transmission speeds in excess of 9600 bits/sec are the maximum that can be carried by analogue transmission systems other than the 48 kbits/sec system outlined earlier. A 24-channel p.c.m. system operates at 1.536 Mbits/sec and a 30-channel p.c.m. system at 2.048 Mbits/sec over unloaded cable with pulse regenerators at 1.828 km intervals. The bit rate allocated per speech channel in either of these systems is 64 kbits/sec which is much larger than the maximum 9600 bits/sec that analogue systems are capable of providing. This means that both a p.c.m. channel and a p.c.m. highway are well able to accommodate a number of high-speed data links. A number of data circuits can be inputted into a data multiplexer to provide a multiplexed bit stream at a bit rate not greater than 2.048 Mbits/sec. Very often the number of data circuits that can be provided in this way is in excess of the demand and then it will be more economic to transmit data over one or more of the time-slots of the 2048 Mbits/sec system. The term **time-slot** refers to the time period occupied by one of the speech channels. Each time-slot contains 8 bits. The two methods are illustrated by Figs. 10.17*a* and *b*. In both cases, a number of practical difficulties, beyond the scope of this book, exist in connection with the necessary timing and synchronization arrangements.

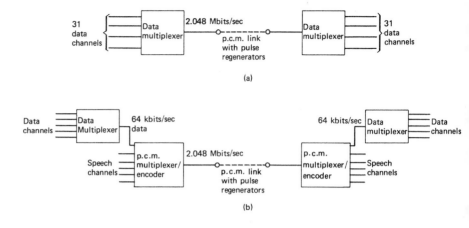

Fig. 10.17 Two methods of transmitting data over a p.c.m. link

The use of a digital circuit to carry data signals offers a number of advantages:

(1) Impulse noise is eliminated.
(2) White noise is not cumulative.
(3) Because of (1) and (2) data can be transmitted at higher bit rates.
(4) An expensive modem is not needed at each end of the data link.

The main disadvantage associated with the transmission of data over a p.c.m. link is concerned with the necessary synchronization between the transmitter and the receiver. The pulse regenerators fitted at regular intervals along a p.c.m. route use timing information derived from the incoming bit stream and this method works very well when speech is transmitted. Data signals could well contain long sequences of 1s or 0s that could cause the regenerator timing to be lost unless the HDB3 code is used.

Exercises

10.1. (i) Write down the numbers 17, 110, and 225 in binary form. (ii) The number 36 is to be transmitted using a 7-bit binary code. Draw the waveform of the signal sent to line using (*a*) unipolar binary, (*b*) a.m.i. code, (*c*) HDB3 code.

10.2. Discuss the factors which influence the maximum speed with which digital signals can be transmitted over a link. Explain why a p.c.m. system transmitting 30 speech channels, each 4 kHz wide, will occupy a much wider bandwidth than if the same number of channels were to be transmitted using frequency division multiplex.

10.3. Explain the meanings of the following terms used in connection with p.c.m.: (i) sampling, (ii) quantization. Calculate the bandwidth occupied by a p.c.m. system if there are 24 channels, each 4 kHz wide, and each sampled amplitude is coded into one of 128 quantization levels.

10.4. (i) Draw and explain the waveform of a pulse amplitude modulated wave.

(ii) Explain with the aid of a block diagram how a p.a.m. signal is converted into p.c.m.

10.5. Explain what is meant by non-linear quantization in a p.c.m. system and why it is commonly employed. What is meant by the term quantization noise?

10.6. Explain the meanings of the terms quantization and quantization noise. By means of a waveform diagram show how the quantized signal is related to the original signal and how the quantization error waveform can be deduced.

10.7. Give three advantages of the use of a digital network for the transmission of data signals. Explain why a modem is not needed and why this is another advantage of digital transmission. Draw the block diagram of a pulse regenerator and briefly explain how it works.

10.8. State the minimum sampling rate that must be used in a pulse modulation system if the analogue signal is to be reconstructed. Explain why there is a need for synchronization in a pulse system.

Describe, with the aid of waveform diagrams, the principles of operation of a pulse amplitude modulation system.

10.9. Draw the block diagram of the 30-channel p.c.m. system and explain how it works. State the values for (i) the number of sampling levels, (ii) the number of binary pulses transmitted per sample, and (iii) the sampling frequency. Use these figures to calculate (*a*) the line bit rate, (*b*) the necessary bandwidth.

Short Exercises

10.10. Define bit rate. Explain briefly why a faster bit rate is obtained when the noise level is reduced.

10.11. Explain why a modem is not required when data is transmitted over a digital circuit.

10.12. Calculate the bandwidth needed for a p.c.m. system which employs 256 quantization levels.

10.13. A 24-channel p.c.m. system uses a sampling frequency of 10 kHz and 8 bits are transmitted per sample for each channel. Determine the bit rate transmitted to the line.

10.14. Explain the difference between analogue and digital pulse modulation.

10.15. Calculate the number of bits per second transmitted for a 4-channel p.c.m. system if each channel includes frequencies up to 3.4 kHz and 128 quantization levels are used.

10.16. Determine the bandwidth required for a p.c.m. system using 512 sampling levels. Choose suitable values for any other parameters needed.

10.17. Calculate the bandwidth needed for a p.c.m. system using 128 sampling levels if (i) each coded sample is accompanied by synchronization information that occupies the time period of one binary pulse, (ii) the signal is a 4800 bits/sec data waveform.

10.18. Explain briefly why the transmitting and receiving terminals of a p.c.m. system must be accurately synchronized.

11 Optical Fibres

Visible and infra-red light extends over a range of wavelengths of about 0.4 μm to about 1 mm but the use of fibre optics is, at present, restricted to the approximate waveband of 0.8 μm to 1.6 μm. The energy carried by light waves in this waveband can be transmitted by glass fibres which act as *dielectric waveguides*. Essentially, an **optical fibre** consists of a cylindrical glass *core* that is surrounded by a glass *cladding*. The use of an optical fibre to transmit light energy offers a number of advantages over the more conventional transmission techniques using copper or aluminium conductors. These advantages are as follows:

(*a*) Light weight, small-dimensioned cables.
(*b*) Very wide bandwidth.
(*c*) Freedom from electromagnetic interference.
(*d*) Low attenuation.
(*e*) High reliability and long life.

Optical fibre, and the associated light sources, is particularly suited to the transmission of digital signals, and the most important telecommunications applications are in connection with pulse code modulated (p.c.m.) multi-channel telephone systems. Many other applications of fibre optics exist in other fields; for example, automobile electronics and industrial control systems.

Propagation in an Optical Fibre

When a light wave travelling in one medium passes into another medium, its direction of travel will probably be changed. The light wave is said to be *refracted*. The ratio

$$\frac{\text{Sine of angle of incidence}}{\text{Sine of angle of refraction}} = \frac{\text{sine } \varphi_i}{\text{sine } \varphi_r} \quad \text{(see Fig. 11.1)}$$

is a constant for a given pair of media. This constant is known as the **refractive index** η for the two media. If one of the two media is air, the *absolute refractive index* of the other medium is obtained. If the absolute refractive indices of the two media are η_1 and η_2 respectively, then

$$\sin \varphi_i / \sin \varphi_r = \eta_2 / \eta_1$$

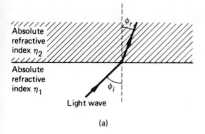

(a)

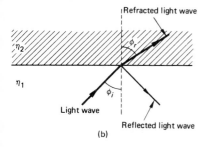

(b)

Fig. 11.1 Reflection and refraction at the boundary of two media with differing refractive indices

When the light passes from a medium of higher refractive index into a medium of lower refractive index, i.e. when $\eta_2 > \eta_1$, the wave will be bent towards the normal (Fig. 11.1a). Conversely, if $\eta_1 > \eta_2$, the light wave will be refracted away from the normal (see Fig. 11.1b). Some of the incident light energy will be reflected at the boundary of the two media. As the angle of incidence is increased, the angle of refraction will also be increased and the point will be reached at which $\varphi_r = 90°$. When this occurs, the light wave will be **totally reflected**. The angle of incidence at which total reflection first occurs is known as the *critical angle* φ_{crit} (see Fig. 11.2). The value of φ_{crit} depends on the absolute refractive indices of the two media according to equation

$$\varphi_{crit} = \sqrt{\frac{2(\eta_1 - \eta_2)}{\eta_1}} \qquad (11.1)$$

The angle of reflection is always equal to the angle of incidence.

Suppose that now another medium, also of refractive index η_2, is placed on the other side of the lower medium, as shown by Fig. 11.2. Provided the angle of incidence φ_1 of the input light wave is larger than the critical value φ_{crit}, the light wave will be able to propagate along the inner medium by means of a series of total reflections. Any other light waves that are also incident on the upper boundary at an angle $\varphi_1 > \varphi_{crit}$ will also propagate along the inner medium. Conversely, any light wave that is incident upon the upper boundary with $\varphi_1 < \varphi_{crit}$ will pass into the upper medium and there be lost by *scattering* and/or *absorption*.

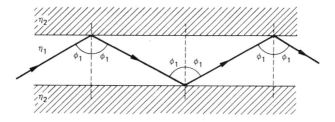

Fig. 11.2 Light wave propagating by multiple reflections

When a number of light waves enter the system and are incident upon the upper boundary with differing angles of incidence, but all greater than φ_{crit}, a number of **modes** are able to propagate. Multimode propagation is shown by Fig. 11.3. It is evident that there are a number of different paths over which the input light waves are able to propagate. These paths are of

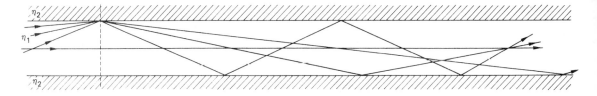

Fig. 11.3 Multimode propagation in a stepped-index fibre

varying lengths and therefore the light waves will take different times to pass over a given distance. This effect is known as *transit time dispersion* and it places a restriction on the maximum possible bit rate that can be transmitted.

Suppose, for example, that a light wave following the direct path takes 5 μs to travel a certain distance, and that a wave travelling over the longest path takes 6 μs to reach the same point. The effect on a rectangular pulse transmitted over the system is shown by Fig. 11.4. The trailing edge of the pulse is delayed in time by 1 μs. This effect will, of course, limit the maximum bit rate that is possible, since the leading edge of a following pulse must not arrive before the extended edge of the pulse shown has ended.

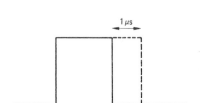

Fig. 11.4 Effect of transit time dispersion on a rectangular pulse

Multimode propagation in an optical fibre can be minimized in one of two ways:

1 A **graded-index fibre** can be used. The refractive index of the inner region or *core* is highest at the centre and then decreases parabolically towards the edges. This ensures that the difference between the refractive indices of the core and the outer medium or cladding is small, and the change can take place smoothly instead of abruptly as before. Light waves nearing the boundary of the two media will then be gradually refracted, rather than reflected, from the boundary as shown by Fig. 11.5. A wave entering the core at a large angle from the horizontal will penetrate a long way from the horizontal before it is refracted sufficiently to change its direction of travel. A wave entering the core at a shallower angle does not penetrate

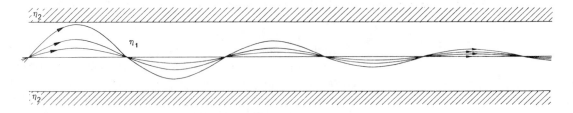

Fig. 11.5 Multimode propagation in a graded-index fibre

as far from the horizontal before it is refracted sufficiently to change its direction of travel.

Once a wave has been refracted back to the horizontal, it will enter the next section of the core at a shallower angle than before and it will not, therefore, travel as far from the horizontal before reversing its direction of travel. The effect of this is to reduce the differences between the lengths of the paths followed by the various light waves. In turn, this ensures that all the light waves travelling through the core have almost the same transit times.

2   A **monomode fibre** can be used. If the diameter of the inner medium or core is reduced to be of the same order of magnitude as the wavelength of the incident light wave, then only one mode will be able to propagate (see Fig. 11.6).

No matter which of the three possible modes of propagation is used, the dimensions of the outer medium or cladding must be at least several wavelengths. Otherwise some light energy will be able to escape from the system, and extra losses will be caused by any light scattering and/or absorbing objects in the vicinity.

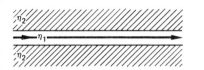

Fig. 11.6 Monomode propagation in a stepped-index fibre

Fibre Optic Cable

An optical fibre cable consists of a glass core that is completely surrounded by a glass cladding. The **core** performs the function of transmitting the light wave(s), while the purpose of the **cladding** is to minimize surface losses and to guide the light waves. The glass used for both the core and the cladding must be of very high purity since any impurities that are present will cause some scattering of light to occur. Two types of glass are commonly employed: silical-based glass (silica with some added oxide) and multi-component glass (e.g. sodium borosilicate)

There are three basic versions of an optical fibre cable.

1   Stepped-index Multimode   The basic construction of a stepped-index multimode optical fibre is shown by Fig. 11.7a and its refractive index profile is shown by Fig. 11.7b. It is clear that an abrupt change in the refractive index of the fibre occurs at the core/cladding boundary. The core diameter, $2r_1$, is often some 50–60 μm but in some cases may be up to about 200 μm. The diameter $2r_2$ of the cladding is standardized, whenever possible, at 125 μm.

2   Stepped-index Monomode   Fig. 11.8a shows the basic construction of a stepped-index monomode optical fibre and Fig.

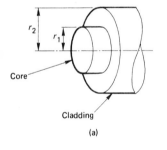

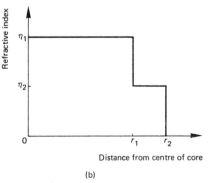

Fig. 11.7 (a) Stepped-index multimode optical fibre, (b) Refractive index profile

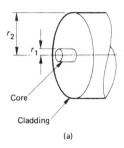

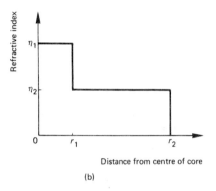

Fig. 11.8 (*a*) Stepped-index mono-mode optical fibre, (*b*) Refractive index profile

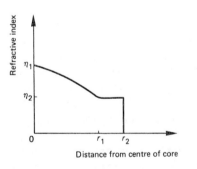

Fig. 11.9 Refractive index profile of graded-index multimode optical fibre

11.8*b* shows its refractive index profile. Once again the change in the refractive indices of the core and the cladding is an abrupt one but now the dimensions of the core are much smaller. The diameter of the core should be of the same order of magnitude as the wavelength of the light to be propagated and it is therefore in the range 1–10 μm. The cladding diameter is the standardized figure of 125 μm.

3 Graded-index Multimode The basic construction of a graded-index multimode optical fibre is the same as that of the stepped-index multimode fibre shown in Fig. 11.7*a*. The refractive index profile is not the same however and this is given by Fig. 11.9. The core has its greatest value of refractive index at its centre and this decreases parabolically towards the boundary with the cladding. The core diameter is in the range 50–60 μm and the cladding diameter is 125 μm.

The relative merits of the three kinds of optical fibre are as follows.

The stepped-index multimode fibre produces large transit time dispersion and in consequence it can only be employed for *bandwidth-distance products* of up to about 50 MHz-km. The use of this kind of optical fibre is therefore restricted to applications such as low-speed data signals and various industrial control systems.

Stepped-index monomode optical fibre can provide very large bandwidth-distance products, of up to about 100 GHz-km, but it is difficult and hence expensive to manufacture and to produce coupling devices to join different lengths of cable.

Graded-index multimode optical fibre has the ability to give bandwidth-distance products of 1 GHz-km or more and, because of its larger dimensions, it is relatively easy to manufacture and instal.

When the required bandwidth is high, as is the case when t.d.m.-p.c.m. is to be transmitted, either stepped-index monomode or graded-index multimode optical fibre is used. The actual bandwidth obtainable is limited by dispersion in the core material, by multipath dispersion in the graded-index fibre, and by the limitations of the optical source and detector. Because of the difficulties associated with the very small core dimensions of the stepped-index monomode fibre, most of the telecommunication systems presently installed and planned use *graded-index multimode fibre*.

Glass is a brittle material and an optical fibre cable must be protected against breakage both during its installation and over its anticipated lifetime. The necessary strength is provided by a steel wire situated in the middle of the cable and protection is given by a plastic sheath. The constructional detail of one type

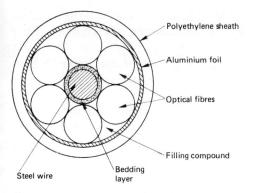

Fig. 11.10 Optical fibre cable

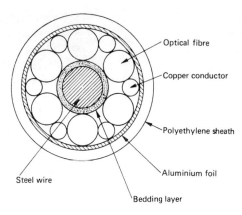

Fig. 11.11 Optical fibre cable with added copper conductors

of cable is shown by Fig. 11.10. The steel wire at the centre of the cable has a bedding layer around it and then the six optical fibres are laid helically around the layer. The gaps between the fibres are filled up with a filling compound and then an aluminium-tape water-barrier is wrapped around to form the cable. Lastly, a polyethylene sheath is placed around the aluminium tape to complete the cable.

In some cases copper conductors are also provided within the cable; they may be used for power-carrying purposes and for speaker circuits. An example of such a cable is shown in Fig. 11.11.

Attenuation of an Optical Fibre

The light power output of the light source has to feed into the optical fibre with the maximum possible efficiency. To this end, *couplers* have been designed to maximize the *launching efficiency*. The launching efficiency is the ratio

$$\eta = \frac{\text{Power accepted by fibre}}{\text{Power emitted by light source}} \times 100\% \qquad (11.2)$$

The design and manufacture of suitable coupling devices is easier and cheaper for graded-index multimode fibre than for stepped-index fibre because of the very small dimensions of the latter.

The light energy fed into an optical fibre is attenuated as it travels towards the far end. The losses in an optical fibre are contributed to by a number of sources. The sources of loss are: absorption; scattering in the core because of inhomogeneities in the refractive index—this is known as Rayleigh scattering;

scattering at the core/cladding boundary; losses at the coupling devices used; and losses due to radiation at each bend in the fibre.

The attenuation coefficient of an optical fibre refers only to losses in the fibre itself, i.e. neglecting coupling and bending losses.

The variation with frequency of the attenuation coefficient of an optical fibre depends upon the specific glass used for the core and the cladding. However, certain features are common to all types of fibre. The loss is high at about 1.7×10^{14} Hz and decreases rapidly with increase in frequency to reach a minimum at about 1.9×10^{14} Hz. Further increase in the frequency results in a gradual increase in the attenuation coefficient up to about 5×10^{14} Hz. However, some sharp peaks in the attenuation coefficient exist at approximately 2.16×10^{14} Hz and 2.42×10^{14} Hz.

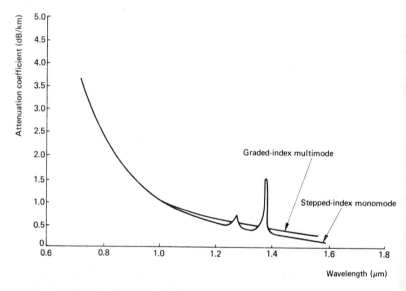

Fig. 11.12 Attenuation/frequency characteristics of stepped-index monomode and graded-index multimode optical fibres

Because of the very high frequencies involved it is customary to work in terms of wavelength rather than frequency. Fig. 11.12 shows typical attenuation/wavelength curves for both stepped-index monomode and graded-index multimode optical fibres. The peaks at 1.24 μm and 1.39 μm occur because of excess absorption loss at these wavelengths.

Optical Sources

The light source used in an optical fibre system must, of course, be able to deliver light energy at the appropriate wavelength. It must also be capable of being modulated at the very high bit rates used, with a low driving power and a relatively high output power. There are two kinds of light source which are able to satisfy these requirements; these are the *light emitting diode* and the *laser diode*.

The light emitting diode (l.e.d.) is generally used for short distance, narrow bandwidth systems, while the laser diode finds application in long-haul wide-bandwidth systems. The laser diode can develop about 100 times more output power than the l.e.d. can provide but it is both more expensive and less reliable at the present time.

The Light Emitting Diode

A semiconductor diode, fabricated using materials such as gallium arsenide (GaAs), gallium aluminium arsenide (GaAlAs), gallium indium arsenide phosphide (GaInAsP) or gallium indium arsenide (GaInAs), will emit light energy whenever it is forward biased and conducting a current. The emitted light may be in either the visible or the infra-red parts of the electromagnetic spectrum. The specific wavelength at which a **light emitting diode** radiates energy can be selected by the suitable choice of the semiconductor material (see Table 11.1).

Table 11.1

Material	GaAlAs	GaAs	GaInAs	GaInAsP
Wavelength of emitted light μm	0.8–0.9	0.9–1.0	1.0–1.1	1.25–1.35

There are two main types of l.e.d. available, one of which radiates the light energy from the surface of the device whilst the other radiates from the edge of the semiconductor structure. The l.e.d. is generally used as the light source for telecommunication systems and produces non-coherent light output with a power of some 0.05 mW to 1 mW. Because of its non-coherent light output, the l.e.d. will emit light waves into an optical fibre at a variety of angles and it is therefore suitable for use with graded-index multimode fibres.

The Laser Diode

The construction of a **laser diode** differs from that of a l.e.d. mainly in that it incorporates a cavity which is resonant at the required frequency of the emitted light. When a voltage is applied across a laser diode, it radiates *coherent* light energy.

Commonly, the semiconductor material employed is gallium aluminium arsenide and this gives a peak emission at a wavelength of 0.82 μm with more than 100 MHz bandwidth. The power output is in the range of 1 mW to 10 mW. Some commercially available laser diodes incorporate optical negative feedback to ensure a stable output over a wide range of ambient temperatures.

The laser diode offers the following advantages over the l.e.d.: (*a*) small dimensions, (*b*) high efficiency, and (*c*) it is easy to modulate.

The laser diode finds its main applications in stepped-index monomode fibre systems operating mainly at a wavelength of 0.85 μm but also at 1.3 μm and at 1.55 μm.

Optical Detectors

The function of an **optical detector**, or photo-detector, is to convert input light energy into the corresponding electrical signal. The detector should have its maximum detection efficiency at the operating wavelength of the system and should operate linearly at the modulated bit rate. Also, of course, it should be of physically small dimensions, cheap and reliable. All of these requirements can be satisfied by the semiconductor **photodiode.**

When a reverse-biased p-n junction is illuminated by light energy, a current will flow across the junction, whose magnitude is very nearly directly proportional to the illumination intensity. The current flowing is very nearly independent of the actual bias voltage so long as it keeps the junction reverse biased. If the diode current is passed through a load resistance, a detected output voltage can be obtained.

Two kinds of photodiode are employed: the silicon *pin diode* and the silicon *avalanche diode*. The pin diode is so named because it has a region of intrinsic semiconductor [EII] sandwiched in between n-type and p-type regions. This device is suitable for the shorter wavelengths but, for wavelengths longer than about 1.1 μm, its sensitivity is no longer good enough and then either a GaInAs or a GaInAsP device would be employed instead. The silicon avalanche diode is essentially a pin device also but it is fabricated in such a way as to accentuate the *avalanche effect* [EII]. The avalanche photodiode can provide a large current gain as well as detection but it has a tendency to be rather noisy.

Optical Fibre Telecommunication Systems

The basic block diagram of an optical fibre telecommunication

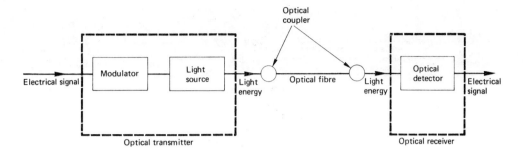

Fig. 11.13 Basic block diagram of an optical fibre system

system is shown by Fig. 11.13. The electrical signal to be transmitted over the system is fed into the optical transmitter. Here it is applied to the modulator and modulates the light source. The modulated light energy is coupled to the optical fibre and is then transmitted to the far end using one of the modes of propagation described earlier. At the optical receiver, incoming light energy is detected in order to convert the signal back to its original electrical format.

The characteristics of the light source make it best suited to some form of digital modulation in which the source will be switched ON and OFF. Some slow-speed systems have been produced in which the input electrical signal is a d.c. data signal. The potential bandwidth offered by an optical fibre is very wide however, and usually the input electrical signal consists of the output from a p.c.m.-t.d.m. system. For example, 120 channels can be transmitted at a bit rate of 8 Mbits/sec or 1920 channels can be transmitted at 140 Mbits/sec. Most such systems operate at a wavelength in the neighbourhood of 0.84 μm but there are also some 1.55 μm systems in use.

The HDB3 coded output signal of the p.c.m. system is not suitable for transmission over an optical fibre and must be converted into a more suitable format before transmission. Fig. 11.14 shows the block diagram of a suitable arrangement. The pure binary coded output of the HDB3 binary converter is scrambled to produce a purely random pulse train and thus avoid any possibility of repetitive patterns occurring (these would tend to cause interference with adjacent optical fibres).

The light source will be either a l.e.d. or a laser diode, while the optical detector is either a pin diode or an avalanche diode. Popular combinations of source/detector are l.e.d./pin-f.e.t. operating at a wavelength of 1.3 μm over graded-index fibre, and laser diode/avalanche diode at 0.85 μm, or laser diode/pin-f.e.t. at either 1.3 μm or 1.5 μm over stepped-index monomode fibre. [Pin-f.e.t. stands for the combination of a pin diode and a gallium arsenide f.e.t.]

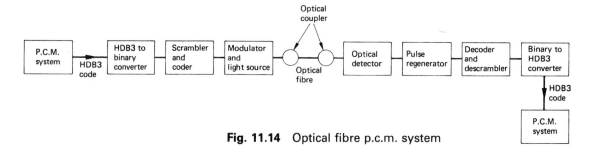

Fig. 11.14 Optical fibre p.c.m. system

The propagating light energy is attenuated as it travels and it may, or may not, be necessary to provide one or more *optical repeaters*. Each optical repeater is able to receive and then regenerate the incoming pulses of light energy, and then re-transmit them into the next section of optical fibre cable. Different kinds of repeater are possible but the most likely consists of a photodiode detector, a pulse regenerator, and then either a l.e.d. or a laser diode.

Exercises

11.1. The repeaters in a conventional coaxial cable system need to be spaced at approximately 2 km intervals along a route. The repeaters in an optical fibre system are spaced at about 10 km intervals. List, and explain, the advantages arising from this greater repeater spacing.

11.2. Describe how light energy can be propagated along an optical fibre by means of reflection and refraction. If, for a particular fibre, the cladding has a refractive index 1% less than the refractive index of the core, calculate the critical angle of incidence.

11.3. Explain, with the aid of a sketch, what is meant by multimode propagation in a stepped-index fibre. How can this effect lead to transit time dispersion? Determine the maximum bit rate possible if 1 μs wide pulses are used and if the transit time of the longest path is 1 μs longer than that of the direct path.

11.4. Draw the block diagram of an optical fibre transmission system. Describe the function of each of the blocks drawn. State what kind of modulation is employed.

11.5. List the sources of loss in an optical fibre system. Briefly explain each of them. Draw a typical attenuation/wavelength characteristic and say why particular wavelengths only are so far used for optical fibre systems.

Short Exercises

11.6. The operating wavelength of an optical fibre system is 0.85 μm. Determine the dimensions of the cable cores shown in Figs. 11.7a and 11.8a in terms of wavelengths.

11.7. An optical fibre cable has a bandwidth-distance product of 2 GHz-km. Determine (i) the maximum frequency that can be used over a distance of 20 km, (ii) the maximum cable length possible for transmission at 500 MHz.

11.8. Explain the terms stepped-index and graded-index when applied to an optical fibre.

11.9. Draw the refractive index profiles of stepped-index multimode, stepped-index monomode, and graded-index multimode optical fibres. Briefly explain how the last *two* reduce transit time dispersion.

11.10. List the devices used as (i) light sources, (ii) light detectors. State their relative merits.

11.11. Some optical fibre telecommunication systems use a light-source/light-detector combination known as l.e.d./pin-f.e.t. State the meaning of these initials and outline the advantages of this choice of devices.

11.12. Describe, with the aid of sketches, the construction of a typical optical fibre cable. Name two types of glass that are commonly used for the core and for the cladding.

Appendix 1
T.T.L. Family

(Includes gates, multivibrators and counters)
(o.c. denotes open collector)
7 400 quad 2-input NAND gates
7 401 quad 2-input NAND gates o.c.
7 402 quad 2-input NOR gates
7 403 quad 2-input NOR gates o.c.
7 404 hex inverters
7 405 hex inverters o.c.
7 406 6 inverter drivers o.c.
7 407 6 drivers o.c.
7 408 quad 2-input AND gates
7 409 quad 2-input AND o.c.
7 410 triple 3-input NAND gates
7 411 triple 3-input AND gates
7 412 triple 3-input NAND gates o.c.
7 413 dual 4-input NAND Schmitt
7 415 triple 3-input AND gates o.c.
7 416 6 inverter drivers o.c.
7 417 6 drivers o.c.
7 420 dual 4-input NAND gates
7 421 dual 4-input AND gates
7 422 dual 4-input NAND o.c.
7 426 quad 2-input interface NAND gates
7 427 triple 3-input NOR gates
7 428 quad 2-input NOR buffers
7 430 8-input NAND gate
7 432 quad 2-input OR gates
7 433 quad 2-input NOR buffers o.c.
7 437 quad 2-input NAND buffers
7 438 quad 2-input NAND buffers o.c.
7 440 dual 4-input NAND buffers
7 450 dual exp. 2-wide 2-input A.O.I gates
7 451 dual 2-wide 2-input A.O.I gates
7 453 exp. 4-wide 2-input A.O.I gates
7 454 4-wide A.O.I gate
7 455 2-wide 4-input A.O.I gate
7 470 edge-triggered J-K flip-flop
7 471 gated R-S master-slave flip-flop
7 472 gated J-K master-slave flip-flop
7 473 dual J-K master-slave flip-flop
7 474 dual D positive edge-triggered flip-flop
7 475 quad D latches
7 476 dual J-K master-slave flip-flops

7 478 dual J-K master-slave flip-flops
7 486 quad 2-input exclusive OR gates
7 490 decade counter
7 492 divide-by-2 and divide-by-6 counter
7 493 4-bit binary counter
7 4107 dual J-K master-slave flip-flops
7 4109 dual J-K positive edge-triggered flip-flops
7 4112/3/4 dual J-K negative edge-triggered flip-flops
7 4121 monostable multivibrator
7 4122 retriggerable monostable multivibrator
7 4136 quad exclusive-OR gates o.c.
7 4160 4-bit synchronous decade counter with clear
7 4161 4-bit synchronous binary counter with clear
7 4162 4-bit synchronous decade counter
7 4163 4-bit synchronous binary counter
74 174 hex D flip-flop
74 175 quad D flip-flop
74 176 preset decade and binary counter
74 177 preset decade and binary counter
74 190 synchronous up-down b.c.d. counter
74 191 synchronous up-down binary counter
74 192 synchronous up-down decade counter
74 193 synchronous up-down 4-bit counter
74 260 dual 5-input NOR gates
74 266 quad 2-input exclusive-OR o.c.
74 290 decade counter
74 293 4-bit binary counter
74 386 quad 2-input exclusive-OR gates

Appendix 2
C.M.O.S. Family

(Includes gates, multivibrators, and counters)
4 000 dual 3-input NOR gates plus an inverter
4 001 quad 2-input NOR gates
4 002 dual 4-input NOR gates
4 011 quad 2-input NAND gates
4 013 dual D-type flip-flops
4 016 analogue gate
4 017 decade counter
4 018 presettable divide-by-n counter
4 019 quad AND-OR select gate
4 022 divide-by-8 counter
4 023 triple 3-input NAND gate
4 024 7-stage binary counter
4 025 triple 3-input NAND gate
4 026 decade counter
4 027 dual J-K flip-flop
4 029 presettable up-down counter
4 030 quad exclusive-OR gates
4 042 quad D-type clocked latch
4 043 quad 3-state NOR S-R latch
4 044 quad 3-state NAND S-R latch
4 047 monostable/astable multivibrator
4 068 8-input NAND gate
4 070 quad exclusive-OR gates
4 071 quad 2-input OR gates
4 072 dual 4-input OR gates
4 073 triple 3-input AND gates
4 075 triple 3-input OR gates
4 076 quad D-type flip-flop
4 077 quad exclusive-NOR gates
4 081 quad 2-input AND gates
4 082 dual 4-input AND gates
4 096 gated J-K flip-flop
4 510 bcd up-down counter
4 516 binary up-down counter
4 528 dual retriggerable monostable multivibrators

Appendix 3
Binary Arithmetic

Table 1

2^7	2^6	2^5	2^4	2^3	2^2	2^1	2^0
128	64	32	16	8	4	2	1

The value of each power of 2 is given in the table and any desired number can be obtained by the correct choice of 0s and 1s. Thus the number 21, for example, is equal to $16 + 4 + 1$ and it is therefore given by

0 0 0 1 0 1 0 1 in the 8-unit binary code

or by

1 0 1 0 1 if only 5 bits are used

Some other binary equivalents of denary numbers are given in Table 2, a 7-unit code being assumed.

In digital electronic systems, the active devices employed are operated as switches and have two stable states, ON and OFF. For this reason, the binary numbering system is used, in which only two digits 1 and 0 are allowable. Larger numbers are obtained by utilizing the various powers of 2. The least significant bit (l.s.b.) of a binary number represents a multiple (0 or 1) of 2 and is (normally) written at the right-hand side of the number. The next digit to the left represents a multiple of 2 and so on as shown by Table 1.

Table 2

11	0	0	0	0	1	0	1	1
43	0	0	1	0	1	0	1	1
63	0	0	1	1	1	1	1	1
111	0	1	1	0	1	1	1	1

Base or Radix Conversion

(a) Decimal to Binary

To convert a decimal integer number into its binary equivalent, the decimal number should be repeatedly divided by 2, and each time the remainder, which will be either 0 or 1, should be recorded. Eventually the number will be reduced to 1, at which stage further division will not give an integer number, and so the quotient 1 is considered to be a remainder of 1. The required binary number is then obtained by writing down the remainders in reverse order.

EXAMPLE 1

Convert 38 into binary.

Solution

number	38	19	9	4	2	1
remainder		0	1	1	0	0	1

Therefore   38 = 100110

EXAMPLE 2

Convert 277 into binary.

Solution

number	277	138	69	34	17	8	4	2	1
remainder		1	0	1	0	1	0	0	0	1

Therefore   277 = 100010101

(b) Decimal Fractions to Binary Fractions

To convert a decimal fractional number into the corresponding binary fraction, multiply the decimal fraction repeatedly by 2 and each time record the integer number obtained. The required binary fraction is then obtained by reading the integers from left to right.

EXAMPLE 3

Convert 0.426 to binary.

Solution

fraction	0.426	0.852	1.714	1.428	0.856	1.712	1.424
integers		0	1	1	0	1	1	etc.

Therefore   0.426 = 0.011011 etc.

EXAMPLE 4

Convert 0.125 into binary.

Solution

fraction	0.125	0.25	0.5	1.0
integers		0	0	1

Therefore   0.125 = 0.001

(c) Binary to Denary

The conversion of binary numbers into their denary equivalents is best achieved using either Table 1 or Table 3.

Table 3

2^{-1}	2^{-2}	2^{-3}	2^{-4}	2^{-5}
0.5	0.25	0.125	0.0625	0.03125

EXAMPLE 5

Convert 10110.101 into denary.

Solution
Using Tables 1 and 2,
10110.101
$$= 1 \times 16 + 0 \times 8 + 1 \times 4 + 1 \times 2 + 0 \times 1 + 1 \times 0.5 + 0 \times 0.25 + 1 \times 0.125$$
$$= 22.625$$

Arithmetic Operations

The processes of binary addition, subtraction, multiplication and division are essentially the same as in ordinary base 10 arithmetic but are, of course, restricted to the use of the two digits 1 and 0.

(a) Binary Addition

The rules for the addition of binary numbers are given by Table 4.

When two 1s are added together their sum is 10 and so the sum in that order of unit is 0 with a carry of 1.

Table 4

A	B	Sum	Carry
0	0	0	0
1	0	1	0
0	1	1	0
1	1	0	1

EXAMPLE 6

Add the binary numbers 10111 and 01101.

Solution
```
  1 0 1 1 1 = 23
+ 0 1 1 0 1 = 13
 ─────────
1 0 0 1 0 0 = 36
```

(b) Binary Subtraction

The rules for performing binary subtraction are given in Table 5.

When a number is to be subtracted from a smaller number (always $0-1$), a 1 must be borrowed from the next column to the left. This 1 is a power of 2 higher and hence the subtraction becomes $2-1=1$.

Table 5

A	B	Difference	Borrow
0	0	0	0
1	0	1	0
0	1	1	1
1	1	0	0

EXAMPLE 7

Subtract 10101 from 11011.

Solution

$$
\begin{array}{r}
1\,1\,0\,1\,1 = 27 \\
-1\,0\,1\,0\,1 = 21 \\
\hline
0\,0\,1\,1\,0 = 6
\end{array}
$$

EXAMPLE 8

Subtract 111010 from 1011111.

Solution

$$
\begin{array}{r}
1\,0\,1\,1\,1\,1\,1 = 95 \\
-1\,1\,1\,0\,1\,0 = 58 \\
\hline
1\,0\,0\,1\,0\,1 = 37
\end{array}
$$

The subtraction of binary numbers is more easily carried out electronically when an alternative method of subtraction is employed. A binary number has two *complements*, known as the ones complement and the twos complement. The ones complement of a binary number is obtained by changing all the 1s in the number into 0s and all the 0s into 1s. The twos complement is obtained by adding 1 to the 1s complement.

The subtraction of a number x from another number y is achieved in the following way. The ones or the twos complement of x is obtained and is then *added* to y. If the left-hand digit of the sum is 0, the difference is *negative*; conversely, if the left-hand digit is 1, a positive difference has been obtained. This left-hand digit is known as the *sign* digit since it indicates the sign ($\pm$) of the difference.

When the sign digit is 0, the result is negative and the complement of the difference is obtained. When the sign digit is 1, the result of adding y to the complement of x actually is the required difference when the twos complement has been used. When the ones complement has been used, the required difference *minus* 1 is obtained; in this case the positive sign digit 1 must be shifted around to the right-hand side of the number and *added* to the digit already there.

EXAMPLE 9

Subtract 10101 from 11011 using the twos complement method.

Solution

$$\begin{array}{r} 1\,1\,0\,1\,1 = 27 \\ -1\,0\,1\,0\,1 = 21 \end{array} = \begin{array}{r} 1\,1\,0\,1\,1 \\ +\,0\,1\,0\,1\,1 \\ \hline 1\,0\,0\,1\,1\,0 = +6 \end{array} \text{ twos complement}$$

EXAMPLE 10

Subtract 11011 from 10101 using the twos complement method.

Solution

$$\begin{array}{r} 1\,0\,1\,0\,1 = 21 \\ -1\,1\,0\,1\,1 = 27 \end{array} = \begin{array}{r} 1\,0\,1\,0\,1 \\ +\,0\,0\,1\,0\,1 \\ \hline \text{negative sign} \rightarrow 0\,1\,1\,0\,1\,0 \end{array} \text{ twos complement}$$

The ones complement is $11010 - 1$ or 11001 and so the required difference is

$$0\,0\,1\,1\,0 = -6$$

EXAMPLE 11

Subtract 11101 from 01011 using the twos complement method.

Solution

$$\begin{array}{r} 0\,1\,0\,1\,1 = 11 \\ -1\,1\,1\,0\,1 = 29 \end{array} = \begin{array}{r} 0\,1\,0\,1\,1 \\ +\,0\,0\,0\,1\,1 \\ \hline \text{negative sign} \rightarrow 0\,0\,1\,1\,1\,0 \end{array} \text{ twos complement}$$

The ones complement is 01101 and so the required difference is

$$10010 = -18$$

EXAMPLE 12

Subtract 11101 from 01011 using the ones complement method.

Solution

$$\begin{array}{r} 0\,1\,0\,1\,1 = 11 \\ -1\,1\,1\,0\,1 = 29 \end{array} = \begin{array}{r} 0\,1\,0\,1\,1 \\ +\,0\,0\,0\,1\,0 \\ \hline 0\,0\,1\,1\,0\,1 \end{array} \text{ ones complement}$$

and so the difference is

$$10010 = -18$$

EXAMPLE 13

Subtract 01011 from 11101 using the ones complement method.

Solution

$$\begin{array}{r} 1\,1\,1\,0\,1 = 29 \\ -0\,1\,0\,1\,1 = 11 \end{array} = \begin{array}{r} 1\,1\,1\,0\,1 \\ +1\,0\,1\,0\,0 \quad \text{ones complement} \\ \hline 1\,1\,0\,0\,0\,1 \end{array}$$

The positive sign digit (1) must now be shifted around to the right-hand side of the number and added to the digit already there. Hence the difference is

$$10010 = 18$$

(c) Binary Multiplication

The product of two binary numbers is 1 only if both the digits are 1. Otherwise it is 0.

EXAMPLE 14

Multiply 11011 by 10101.

Solution

$$\begin{array}{r} 1\,1\,0\,1\,1 = 27 \\ \times 1\,0\,1\,0\,1 = 21 \\ \hline 1\,1\,0\,1\,1 \\ 0\,0\,0\,0\,0 \\ 1\,1\,0\,1\,1 \\ 0\,0\,0\,0\,0 \\ 1\,1\,0\,1\,1 \\ \hline 1\,0\,0\,0\,1\,1\,0\,1\,1\,1 = 1+2+4+16+32+512 \\ = 567 \end{array}$$

(d) Binary Division

Binary division is carried out in the same way as the more customary denary division.

EXAMPLE 15

Divide 11011 by 01001.

Solution

$$\begin{array}{r} 11 \\ 0\,1\,0\,0\,1\,)\overline{1\,1\,0\,1\,1} \\ \underline{1\,0\,0\,1} \\ \hline 1\,0\,0\,1 \\ \underline{1\,0\,0\,1} \qquad = 11 = 3 \\ \hline 0\,0\,0\,0 \end{array}$$

Appendix 4 Multiple Choice Questions

In all of the following questions, *one* of the four given answers is correct. The solutions are given on page 179.

1. Three inputs to a NAND gate are A. B and C. The output of the gate is

(*a*) $A + B + C$

(*b*) $\overline{A \, B \, C}$

(*c*) $\overline{A} \, \overline{B} \, C$

(*d*) $\overline{A} \, \overline{B} + C$

2. A truth table shows

(*a*) The input state of a circuit for all possible combinations of the output state.

(*b*) The output state of a circuit for all possible combinations of input states.

(*c*) The number of inputs which must be at 1 for the output to become 1.

(*d*) The number of inputs which are simultaneously equal to 1.

3. When a gate is enabled,

(*a*) its power supplies are switched on.

(*b*) one of its inputs goes to logical 1.

(*c*) all of its inputs go to logical 1.

(*d*) the input signal is allowed to pass to the output by the application of a control signal to the other input.

4. A logic circuit has its output at logical 1 when any one of its inputs is at 1. The Boolean expression for the circuit operation is

(*a*) $F = A \, B + C$

(*b*) $F = A \, C + B$

(*c*) $F = B \, C + A$

(*d*) $F = A + B + C$

5. The expression $F = S + R \, S$ simplifies to

(*a*) $F = R$

(*b*) $F = R \, S$

(*c*) $F = S + R$

(*d*) $F = S$

6. The expression $F = X(\bar{X} + Y)$ simplifies to

(*a*) $F = X \, Y$

(*b*) $F = \bar{X} \, Y$

(*c*) $F = X + Y$

(*d*) $F = \bar{X} + Y$

7. For the output of the circuit shown in the figure to be 1,

(*a*) $A = 0, B = C = 1$

(*b*) $A = B = 1, C = 0$

(*c*) $A = B = 0, C = 1$

(*d*) $A = 1, B = 0, C = 1$

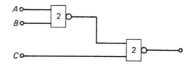

8.

	A		$\bar{A}$		
C	1	0	0	1	D
	1	0	0	1	$\bar{D}$
$\bar{C}$	1	1	1	1	
	1	1	1	1	D
	B	$\bar{B}$	B		

(a) $\bar{F} = A\,B\,\bar{C}$

(b) $\bar{F} = \bar{B}\,\bar{C}\,D$

(c) $\bar{F} = A\,D$

(d) $\bar{F} = \bar{B}\,C$

9. The noise margin of a gate is
(a) The difference between the logic level at the gate output and the threshold value of the gate.
(b) The maximum noise voltage at the gate output.
(c) The minimum noise voltage at the gate output.
(d) The noise level that just causes the gate to mis-operate.

10. The fan-in of a gate is
(a) The minimum number of possible inputs.
(b) The maximum number of possible inputs.
(c) The range of input voltages that will be taken as logical 1.
(d) The range of d.c. supply voltages allowable.

11. The logical function performed by the wired-OR gate is
(a) $\overline{A\,\bar{B} + \bar{C}\,\bar{D}}$
(b) $\overline{A\,B + C\,D}$
(c) $A\,B + \overline{C\,D}$
(d) $A\,B + C\,D$

12. Schottky t.t.l. has the advantage over basic t.t.l. in that
(a) The power dissipation is increased.
(b) The power dissipation is decreased.
(c) The fan-out is increased.
(d) The propagation delay is decreased.

13. Any unused inputs to a t.t.l. gate should
(a) be connected to a used input.
(b) be short-circuited.
(c) be open-circuited.
(d) be connected to earth.

14. The main advantage of the c.m.o.s. family over the t.t.l. family is
(a) Reduced power dissipation.
(b) Increased speed of operation.
(c) Cheapness.
(d) Smaller physical size.

15. The main advantage claimed for the e.c.l. family of logic gates is
(a) The use of a negative power supply voltage.
(b) Low power dissipation.
(c) Large fan-in.
(d) Very low propagation delay.

16. The truth table of an S-R flip-flop is

S	R	Q	Q^+
0	0	0	0
0	0	1	1
1	0	0	1
1	0	1	1
0	1	0	0
0	1	1	0
1	1	0	X
1	1	1	X

(a)

S	R	Q	Q^+
0	0	0	0
0	0	1	1
1	0	0	0
1	0	1	0
0	1	0	0
0	1	1	0
1	1	0	X
1	1	1	X

(b)

S	R	Q	Q^+
0	0	0	0
0	0	1	1
1	0	0	1
1	0	1	1
0	1	0	1
0	1	1	1
1	1	0	X
1	1	1	X

(c)

S	R	Q	Q^+
0	0	0	0
0	0	1	1
1	0	0	1
1	0	1	1
0	1	0	0
0	1	1	0
1	1	0	1
1	1	1	0

(d)

17. A D flip-flop can be made from an S-R flip-flop by
(a) Connecting the R terminal to the S terminal via an inverter.
(b) Using the commoned S and R terminals as the input.
(c) Commoning the Q and the $\bar{Q}$ terminals.
(d) By clocking the flip-flop.

18. A D flip-flop can be used as a divide-by-two circuit by
(a) Connecting its Q output to its D input.
(b) Connecting its $\bar{Q}$ output to its D input.
(c) Connecting its Q and $\bar{Q}$ outputs together.
(d) Connecting its Q output to its D input via an inverter.

19. Master-slave operation of a clocked flip-flop is used
(a) To reduce the power dissipation.
(b) To increase the switching speed.
(c) To ensure reliable switching at the correct times.
(d) To increase the fan-out.

20. A J-K flip-flop will act as a divide-by-two circuit if pulses are applied to the clock input and the J and K terminals are
(a) shorted-circuited.
(b) connected to earth.
(c) connected to logical 1 voltage.
(d) connected to logical 0 voltage.

21. A five-stage binary counter can give a maximum count of
(a) 16
(b) 5
(c) 32
(d) 10

22. In a synchronous counter
    (*a*) All the stages operate at the same time.
    (*b*) The circuitry is simplified.
    (*c*) The stages operate one after the other.
    (*d*) A clock pulse need not be applied.

23. A synchronous counter is faster to operate than a non-synchronous type because
    (*a*) The clock frequency is limited only by the delays of one stage plus the delays of the associated gates.
    (*b*) Four stages are needed for a given count.
    (*c*) Reset terminals can be utilized.
    (*d*) Schottky transistors can be used.

24. A store or memory is said to be volatile if
    (*a*) Its access time is very short.
    (*b*) It is easily affected by external noise sources.
    (*c*) It loses its stored data when its power supplies are turned off.
    (*d*) It retains its stored data when its power supplies are turned off.

25. A ferrite core store is
    (*a*) Fast to access, cheap, and of small physical dimensions.
    (*b*) Slow to access, cheap, and of small physical dimensions.
    (*c*) Slow to access, dear, and occupies a lot of space.
    (*d*) Fast to access, dear, and occupies a lot of space.

26. The hysteresis loop of a ferrite core should be as nearly rectangular as possible to
    (*a*) Increase the magnetic saturation.
    (*b*) Ensure that small changes in m.m.f. do not produce much change in flux density.
    (*c*) Ensure that small changes in m.m.f. produce large changes in flux density.
    (*d*) Increase the permeability of the ferrite.

27. A 30-kilobit store contains
    (*a*) 30 000 cores.
    (*b*) 3000 cores.
    (*c*) 3067 cores.
    (*d*) 30 670 cores.

28. A modem in a data system
    (*a*) Converts the digital data into analogue.
    (*b*) Identifies the distant data terminal.
    (*c*) Amplifies the signal before transmission.
    (*d*) Removes any signalling tones that are present on the line.

29. Frequency shift modulation
    (*a*) Uses two separate frequencies to represent binary 1 and binary 0.
    (*b*) Deviates a carrier frequency above its mean value to indicate binary 1 and leaves the carrier frequency unchanged to indicate binary 0.
    (*c*) Deviates a carrier frequency above and below its mean value to represent binary 1 and binary 0.
    (*d*) Deviates a carrier frequency by 1000 Hz to represent binary 1 and by 2000 Hz to represent binary 0.

30. For a 1200 bits/sec frequency-shift data system, binary 1 and binary 0 are represented, respectively, by
    (*a*) 1300 Hz and 1900 Hz.
    (*b*) 1300 Hz and 2100 Hz.
    (*c*) 1900 Hz and 2100 Hz.
    (*d*) 1300 Hz and 1700 Hz.

31. In a 600 bits/sec data system using frequency-shift modulation, the two frequencies transmitted to line are 1300 Hz and 1700 Hz. The nominal carrier frequency of this system is
(*a*) 1600 Hz
(*b*) 1300 Hz
(*c*) 1900 Hz
(*d*) 1500 Hz.

32. A 1200 bits/sec data system requires a minimum bandwidth of
(*a*) 600 Hz
(*b*) 2400 Hz
(*c*) 1200 Hz
(*d*) 900 Hz.

33. A four-state phase modulation system transmits the binary number 11 10 01 00 10 11. The changes of phase that are signalled to line are
(*a*) 45°, 135°, 90°, 0°, 135°, 45°
(*b*) 315°, 225°, 135°, 45°, 225°, 315°
(*c*) 135°, 45°, 225°, 315°, 45°, 135°
(*d*) 45°, 45°, 45°, 45°, 45°, 45°.

34. A vestigial sideband system is one in which
(*a*) Only one sideband plus a pilot carrier is transmitted.
(*b*) A part of each sideband is transmitted.
(*c*) All of one sideband and a part of the other sideband are transmitted.
(*d*) Both sidebands and a vestige of the carrier are transmitted.

35. Digital signals are changed into analogue form before transmission over the p.s.t.n. because
(*a*) The signal-to-noise ratio is improved.
(*b*) The required bandwidth is minimized.
(*c*) The p.s.t.n. cannot transmit digital signals.
(*d*) Digital signals would falsely operate the telephony signalling equipment.

36. Double-current working of d.c. circuits is more often used than single-current operation because
(*a*) It gives faster and more reliable working.
(*b*) Single-current operation is not possible.
(*c*) A narrower bandwidth is necessary.
(*d*) The waveform of the line current is not dependent upon the speed of signalling.

37. The advantage of voice-frequency operation of a data link over d.c. signalling is
(*a*) Modems can be used.
(*b*) The speed of transmission can be increased.
(*c*) A use is found for spare telephone cables.
(*d*) Data signals can be transmitted by telephone.

38. A p.c.m. system uses 256 sampling levels. The number of binary pulses needed per sample, neglecting synchronization pulses, is
(*a*) 7
(*b*) 8
(*c*) 256
(*d*) 128

39. A signal containing components in the frequency band 300–3000 Hz is transmitted over a p.c.m. link. The minimum sampling frequency necessary is
(*a*) 300 Hz
(*b*) 3000 Hz
(*c*) 600 Hz
(*d*) 6000 Hz

40. The end terminals of a p.c.m. system must be synchronized together so that
(*a*) The pulse waveform is not lost.
(*b*) The quantization is accurate.
(*c*) Quantization noise is minimized.
(*d*) Incoming pulses are directed to their appropriate channels.

41. In a magnetic core memory the purpose of the sense wire is for
(*a*) Writing data into store.
(*b*) Identifying the X or Y co-ordinates.
(*c*) Reading data out of store.
(*d*) Re-writing data into store.

42. If the D input of a D-type flip-flop is held at logical 1 and the clock input goes from 1 to 0, the outputs will
(*a*) Remain the same irrespective of their initial state.
(*b*) Change irrespective of their initial state.
(*c*) Become $Q = 0$, $\bar{Q} = 1$.
(*d*) Become $Q = 1$, $\bar{Q} = 0$.

Digital Techniques III: Learning Objectives (TEC)

DIGITAL TECHNIQUES IIIA

(A) Bistables

(B) Counters

page 63, 64, 66	2.5	Demonstrates using the counter of 2.4 how, with additional gating, the count may be curtailed at any desired value.
65	2.6	Identifies the need for a decade counter and demonstrates the feedback paths needed to convert the counter of 2.4 into a divide-by-ten device.
73	2.7	Compares TTL and CMOS counter circuits and devices in terms of their speed of operation and output configuration.
73	2.8	States that CMOS counting devices of up to 21 stages are currently available and quotes their limitations in terms of input frequency and availability of outputs.
73	2.9	Describes an application of the device in 2.8.

(C) Registers

77	(3)	*Knows that bistable elements may be used to form registers and appreciates the operation and applications of these devices.*
77	3.1	Sketches the simple logic diagram of a four bit register showing (*a*) Serial input (*b*) Parallel inputs (*c*) Clock input (*d*) Outputs.
78, 79	3.2	Describes the operation of the register in 3.1 and describes one application of the circuit.
79	3.3	Sketches the logic diagram of a four bit register which incorporates the following facilities: (*a*) Serial inputs (*b*) Parallel inputs (*c*) Clock input (*d*) Shift mode control (*e*) Left and right shift capability.
79	3.4	Describes the operation of the circuit of 3.3.
78	3.5	Explains how the four-bit register may be used for the storage and manipulation of digital numbers.
80	3.6	States that registers of both types are commercially available in TTL and CMOS form.
80	3.7	Compares the performance of TTL and CMOS registers in terms of speed and flexibility of operation.

(D) Logic Circuit Families

27	(4)	*Knows the basic circuit configurations and characteristics of the currently available range of commercial logic integrated circuits.*

page 35 4.1 Sketches the circuit of a TTL NAND gate with a totem pole output stage and explains its operation.

39 4.2 Sketches the circuit of an open-collector TTL NAND gate, explains its operation and describes at least one application of the device.

43 4.3 Sketches the circuit of an emitter coupled logic OR/NOR gate and describes its operation.

44 4.4 Compares TTL and ECL gates in terms of supply requirements, power dissipation, compatability and the effect of open circuit inputs.

DIGITAL TECHNIQUES IIIB

(A) Karnaugh Maps

16 (1) *Appreciates the fact that Karnaugh maps may be used as a means of simplifying Boolean expressions with up to three independent variables.*

16 1.1 Constructs a Karnaugh map for any expression containing up to three independent variables.

17 1.2 Using either a truth table or a Boolean expression, completes a Karnaugh map for a given function with up to three independent variables.

17 1.3 Using the completed map of 1.2 produces the minimized expression for the given function.

18 1.4 Demonstrates the circuit of the minimized expression of 1.3.

(B) Universal NAND and NOR Logic

9, 16 (2) *Understands the minimization of Boolean expressions.*

12 2.1 Demonstrates, using truth tables to illustrate, that a NAND or NOR gate may be used as an inverter or NOT gate.

9, 12, 14 2.2 States that the rules governing the application of DeMorgan's theorem are
(*a*) Invert the variables
(*b*) Change the connectives
(*c*) Invert the whole expression.

9 2.3 States that since De Morgan's theorem produces the equivalent of any expression to which it is applied it is possible to apply it to any part of, or the whole expression.

11 2.4 Proves the statement of 2.3 by practical means.

13 2.5 Applies De Morgan's theorem to the minimized expressions of 1.3 to obtain circuits in NAND and NOR form.

Transmission Systems III: Learning Objectives (TEC)

(A) Characteristics of a Network

(1) *Understands the effect of cables on analogue and digital signals.*

    1.1   Sketches typical attenuation/frequency and group delay/frequency curves of unloaded and loaded audio cables and coaxial cables.

    1.2   Explains the shape of the curves in 1.1 in terms of the parameters of the cable.

    1.3   Explains how attenuation/frequency and delay/frequency characteristics affect analogue bandwidth and digital bit rate.

page 91 (2) *Understands the effect of the cable upon pulses.*

92, 93   2.1   Sketches and explains current/time graphs representing:
- (*a*) single and double current signals in a circuit with no distortion,
- (*b*) the slow growth and decay of receive current in a line having significant capacitance, using single current signals,
- (*c*) the sending end and receiving end current, using double current signals.

94   2.2   Uses the graphs of 2.1*b*) and *c*) to explain how the use of double current working gives more reliable and allows faster, signalling than single current working.

94   2.3   Sketches graphs illustrating the arrival curve at the end of a long circuit.

94   2.4   Explains the use of the arrival curve and its importance in relation to the maximum speed of working.

92, 93   2.5   Sketches simple circuit elements of unidirectional and both-way arrangements for single current and double current working.

106   2.6   Explains the advantages of voice frequency compared with d.c. signalling over long distances.

(B) Digital Networks

(C) Digital Transmission

(D) Modulation Methods for Data Transmission

(E) Optical Fibres

Answers to Exercises

Numerical Answers to Exercises

7.6 114.2×10^3 km/s **7.14** 300 bits/sec, 100 Hz
7.19 600 bits/sec **8.3** 300 Hz, 600 Hz, 1800 Hz **8.9** 0.67
8.12 600 Hz **8.16** 2400 bits/sec, 4800 bits/sec
8.18 1500 Hz, 1700 Hz **9.11** 1320 bits/sec
9.12 110 bits/sec **9.15** 1.08 **10.3** kilobits/sec **10.11** 36 kHz
10.12 1920 kilobits/sec **10.14** 256 kilobits/sec **10.15** 40 kHz
10.16 40 kHz **11.2** 8° **11.3** 250 kilobits/sec

Answers to Multiple Choice Questions

1 *c*	**2** *b*	**3** *d*	**4** *d*	**41** *c*
5 *d*	**6** *a*	**7** *b*	**8** *d*	**42** *d*
9 *a*	**10** *b*	**11** *b*	**12** *d*	
13 *a*	**14** *a*	**15** *d*	**16** *a*	
17 *a*	**18** *b*	**19** *c*	**20** *c*	
21 *c*	**22** *a*	**23** *a*	**24** *c*	
25 *d*	**26** *b*	**27** *d*	**28** *a*	
29 *a*	**30** *b*	**31** *d*	**32** *a*	
33 *b*	**34** *c*	**35** *c*	**36** *a*	
37 *b*	**38** *b*	**39** *d*	**40** *d*	

Self-test questions

2 Electronic gates

AND, OR, NOR, NAND, NOR gates

Mark each of the following statements True or False.

2.1 The Boolean equation for a 3-input NOR gate is $\overline{A\,B\,C}$

2.2 A NAND gate can be connected to act as an inverter but a NOR gate cannot.

2.3 The circuit shown in Fig. 2.1 acts as a 3-input OR gate.

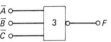

Fig. 2.1

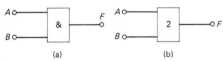

Fig. 2.2

2.4 The BSI symbol for an AND gate is shown in Fig. 2.2*a* but it is allowable to indicate the number of inputs required to be high for the output to be high, i.e. as in Fig. 2.2*b*.

2.5 Table 2.1 is the truth table of an exclusive-OR gate.

Table 2.1

A	0 0 1 1
B	0 1 0 1
F	1 0 1 1

2.6 A 2-input NAND gate can be used as an inverter if its inputs are connected together.

2.7 A 2-input NAND gate can be used as an inverter if one of its imputs is connected to logical 0 voltage level.

2.8 A 2-input NOR gate can be used as an inverter of one of its inputs is connected to logical 0 voltage level.

2.9 If one of the inputs to an AND gate is a control signal it may be used to enable or to inhibit a signal.

2.10 The NAND gate performs the inverse of the OR logical function.

2.11 The NOR gate with inverted inputs performs the AND function.

2.12 The OR gate with inverted inputs performs the NOR function.

2.13 The output of a NAND gate is at logical 1 only when all of its inputs are at logical 0.

2.14 The output of an exclusive-OR gate is at logical 0 when both of its inputs are at logical 0.

2.15 A logic circuit has 5 inputs. The truth table describing the circuit must have 32 rows and 6 columns (or vice versa).

2.16 When a variable is complemented its value is changed from 1 to 0 or from 0 to 1.

2.17 The circuit shown in Fig. 2.3 provides the Boolean equation

$$F = A + B\overline{C}$$

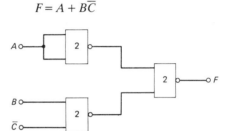

Fig. 2.3

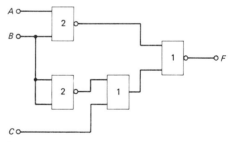

Fig. 2.4

2.18 When $A = B = C = 1$ the output of the circuit given in Fig. 2.4 is 1.

2.19 When $A = B = C = 0$ the output of the circuit given in Fig. 2.4 is 0.

2.20 The output of the circuit shown in Fig. 2.4 is not affected by the logical state of the input C.

2.21 The inputs to a 3-input NAND gate are $\overline{A}$, $\overline{B}$, $\overline{C}$. The output of the gate is $A + B + C$.

2.22 An AND gate with inverted inputs acts as an NOR gate.

2.23 The four inputs to a NOR gate are P, Q, R and S. When $P = Q = R = 0$ and $S = 1$ the output F is 0.

2.24 For the output of the gate in **2.23** to be at logical 1, all four inputs must be at 1.

Multiple Choice Questions: mark the correct answer.

2.25   The BS symbol for an OR gate is shown by *a, b, c* or *d* in Fig. 2.5.

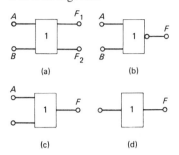

Fig. 2.5

2.26   The symbol shown in Fig. 2.6 represents *a*) an AND gate, *b*) an OR gate, *c*) a NAND gate, or *d*) a NOR gate.

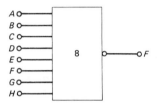

Fig. 2.6

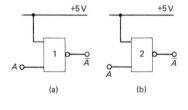

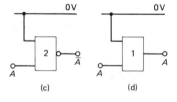

Fig. 2.7

2.27   Which of the circuits shown in Fig. 2.7 is correct?
2.28   Table 2.2 shows the truth table of a 3-input AND gate. If negative logic were to be employed, the circuit would perform the logical function of *a*) OR, *b*) NAND, *c*) NOR, or *d*) NOT.

Table 2.2

A	0 1 0 1 1 0 1 1
B	0 0 1 0 1 0 0 1
C	0 0 0 0 0 1 1 1
F	0 0 0 0 0 0 0 1

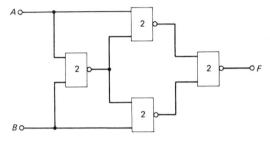

Fig. 2.8

2.29   The circuit shown in Fig. 2.8 implements *a*) the exclusive-OR, *b*) the exclusive-NOR, *c*) the NOR logic function, or *d*) none of these.
2.30   A truth table has 256 columns. How many inputs does the circuit it represents possess: *a*) 256, *b*) 128, *c*) 14, or *d*) 8.
2.31   The AND function can be obtained by the cascade connection of *a*) two NOR gates, *b*) three NAND gates. *c*) three NOR gates, or *d*) two NAND gates.
2.32   The three inputs to a NAND gate are *A, B* and *C*. The output of the gate is

   a) $\overline{AB} + C$, *b*) $\overline{ABC}$, *c*) $\overline{A + B + C}$, or *d*) $\overline{ABC}$

2.33   The output of the circuit given in Fig. 2.9 is

   a) $(A + B)C$, *b*) $BA + C$
   c) $A + B + C$, or *d*) ABC

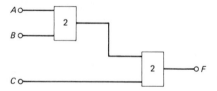

Fig. 2.9

2.34   A logic circuit has *n* input variables. Its truth table must contain *a*) *n*, *b*) $n - 1$, *c*) 2^n, or *d*) $2^n - 1$ entries.
2.35   The logical function $F = A + BC$ is implemented by Fig. 2.10 *a, b, c,* or *d*.

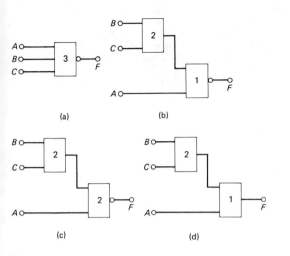

(a)   (b)

(c)   (d)

Fig. 2.10

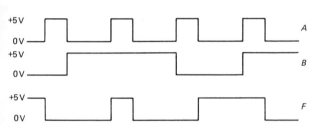

Fig. 2.11

2.36 Fig. 2.11 shows the input and output waveforms of a gate. The gate performs the logical function *a)* AND, *b)* exclusive-NOR, *c)* NOR, or *d)* none of these.
2.37 Fig. 2.12 shows the input and output waveforms of a logical circuit. The circuit performs the function

a) $F = AB + C$, *b)* $A + B + C$
c) $A(B + C)$, or $A + BC$

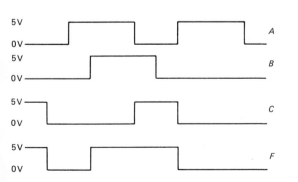

Fig. 2.12

2.38 If negative logic were applied to the waveforms shown in Fig. 2.12 the logical function performed would be

a) BC, *b)* $A + B + C$. *c)* $\overline{A} + \overline{C}$, or *d)* $\overline{A}(B + C)$

Short Exercises
2.39 The waveforms given in Fig. 2.13 are applied to the inputs of a 3-input NAND gate. Sketch the output waveform.

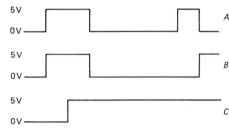

Fig. 2.13

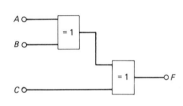

Fig. 2.14

2.40 Obtain an expression for the output F of the circuit given in Fig. 2.14.

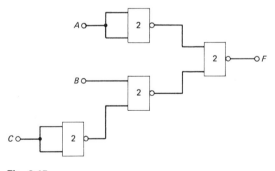

Fig. 2.15

2.41 The waveforms shown in Fig. 2.13 are applied to the input terminals of the circuit of Fig. 2.10*b*. Draw the output waveform.
2.42 Obtain an expression for the output F of the circuit of Fig. 2.15.

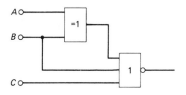

Fig. 2.16

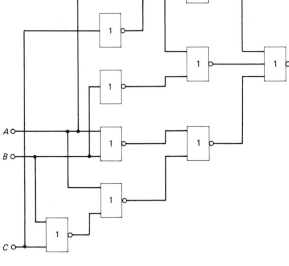

Fig. 2.17

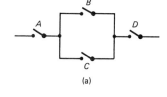

(a)

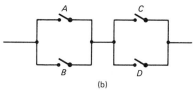

(b)

Fig. 2.18

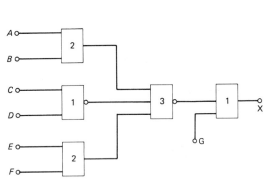

Fig. 2.19

2.43 Write down the truth table for the circuit given in Fig. 2.16.

2.44 Determine the expression for the output F of the circuit in Fig. 2.17.

2.45 Determine the Boolean equation representing each of the circuits shown in Fig. 2.18. Redraw each circuit using AND and/or OR gates.

2.46 In the circuit of Fig. 2.19 the output X is 0. Determine the logical states of each of the inputs A through to G.

Simplification of Boolean equations

Mark each of the following statements True or False.

2.47 The complement of $AB + C$ is $(A + B)C$.

2.48 $(A + B)(A + BC) + \overline{A}\overline{B} + \overline{A}\overline{C} = 1$

2.49 $AB = BA$

2.50 $A + B = B + A$

2.51 $(A + B) + C = C$

2.52 $ABD + BC + A\overline{C} = BC + A\overline{C}$

2.53 If **2.52** is True, then $A\overline{C}$ can be subtracted from both sides.

2.54 $\overline{\overline{AB} + \overline{BC}} = \overline{\overline{AB}} \cdot \overline{\overline{BC}}$

2.55 $\overline{\overline{AB} + B} = 1$

2.56 $\overline{\overline{(A + B)}B} = 1$

2.57 $\overline{A} + \overline{B} = A\overline{B}$

2.58 $ABC + ABD + \overline{A}\overline{B}\overline{D} + \overline{A}D + \overline{\overline{A}\overline{D}} = 1$

2.59 $ABC + \overline{A}BC + A\overline{B}C + AB\overline{C} + \overline{A}\overline{B}C + 1 = 1$

2.60 Only sum-of-product equations can be simplified on a Karnaugh map.

2.61 In a Karnaugh map each square represents A or B or C or D.

2.62 When using a Karnaugh map to simplify an equation, only squares immediately adjacent to one another can be looped.

2.63 In the map shown, the looped squares represents the function BD.

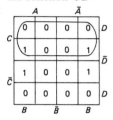

2.64 In the map shown, the looped squares represent the function $\overline{CD}$

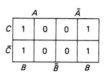

2.65 The mapping for an equation that contains 3 variables has 8 squares.

2.66 In the simplification of an equation using a Karnaugh map, squares can be looped together in twos, threes or fours but not in fives, sixes, or sevens.

2.67 Boolean equations can be simplified using either algebraic or mapping methods.

2.68 If $F = AB + \overline{A}B$ then $\overline{F} = AB + \overline{A}B$.

2.69 If $F = ABC + D$ then $\overline{F} = (\overline{A} + \overline{B} + \overline{C})\overline{D}$.

2.70 If $F = (A + B + C)D$ then $\overline{F} = (\overline{A} + \overline{B} + \overline{C})\overline{D}$.

2.71 If $F = \overline{A}B + \overline{A}B$ then $\overline{F} = AB$.

2.72 $ABC + \overline{A}B\overline{C} + \overline{A}D + AD = 1$

2.73 $\overline{AB} + \overline{CD} + C + \overline{C} = 1$

2.74 $A\overline{B}C + B\overline{C}(\overline{B} + C) + \overline{A}BC = A\overline{B}C$

2.75 $AB + A\overline{B} + \overline{A}B + \overline{A}\,\overline{B} = 1$

2.76 $A(B + 1) + A(A + 1)B = A + B$

2.77 $\overline{\overline{A}(B + \overline{C})} = A + \overline{B}C$

2.78 $\overline{A + BC} = \overline{A}(\overline{B} + \overline{C})$

2.79 $(A + B)(\overline{A}\overline{B}) = A + BC$

2.80 $(ABD + FE) + (\overline{ABD + FE})C = (ABD + FE) + C$

Multiple Choice Questions: mark the correct answer.

2.81 The three inputs to a NAND gate are A, B and C. The output of the gate is
a) $\overline{AB} + C$, b) $\overline{ABC}$, c) $A + B + C$, or d) ABC

2.82 A logic gate has its output at logical 0 when any one or more of its inputs is at logical 1. The Boolean expression describing this circuit is
a) $F = A + B + C$, b) $F = ABC$,
c) $F = \overline{A + B + C}$, or d) $F = \overline{ABC}$

2.83 $(\overline{A}B + C)(A + \overline{B}C)$ reduces to
a) $\overline{A}BC + \overline{B}C$, b) $AC + \overline{B}C$,
c) C, or d) $\overline{A}B + \overline{A}BC + A\overline{C}$

2.84 The function $F = \overline{\overline{AB} + CD} + A\overline{A}$ is equal to
a) 1, b) 0, c) $(A + B)CD$, or d) $BA(\overline{C} + \overline{D})$

2.85 De Morgan's rule states that
a) $\overline{AB} = \overline{A}\,\overline{B}$, b) $\overline{AB} = AB$
c) $\overline{AB} = \overline{A} + \overline{B}$, or d) $\overline{AB} = A + B$

2.86 The other de Morgan rule states that
a) $\overline{A + B} = A + B$, b) $\overline{A + B} = \overline{A} + \overline{B}$
c) $\overline{A + B} = AB$, or d) $\overline{A + B} = \overline{A}\,\overline{B}$

2.87 The mapping shown represents a) A, b) B, c) C, or d) $\overline{A}$.

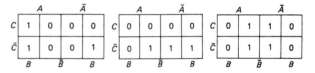

2.88 When the three maps shown are ANDed the function represented is a) $F = 1$, b) $F = 0$, c) $F = AB$, or d) $F = \overline{B}$.

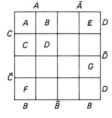

2.89 Referring to the map below left, square A is adjacent to squares a) B, C and D, b) B, C, E, F and G, c) B, C, D, E, and F, or d) B, C, E and F.

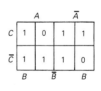

2.90 The minimal solution for the function represented by the map above right is a) $AB + \overline{B}\overline{C} + \overline{A}C$, b) $CB + \overline{A}\overline{B} + A\overline{C}$, c) either a) or b), or d) neither a) nor b).

2.91 $(\overline{A} + B)(A + \overline{B} + C)(\overline{A} + \overline{C}) = (\overline{A} + B)(\overline{B} + \overline{C})$ is a) true, b) false, c) true only if $A = B = 1$, or d) true only if $C = 0$.

2.92 The function $F = (AB + C)(B + CD) + BCD$ is high when
a) $A = B = C = D = 0$, b) $A = C = 0$, $B = D = 1$,
c) $A = B = 0$, $C = D = 1$, or d) $A = B = 1$, $C = D = 0$

Short Exercises

2.93 For the circuit shown in Fig. 2.19 obtain a Boolean expression for the output X of the circuit. Simply the equation if possible and then draw the simplified circuit.

2.94 Obtain an expression for the output F of the circuit given in Fig. 2.20. Simplify the circuit and draw the simpler alternative.

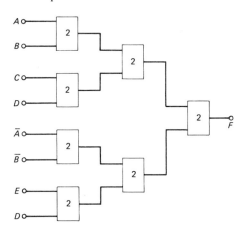

Fig. 2.20

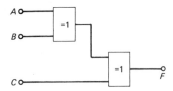

Fig. 2.21

2.95 Derive an expression for the output of the circuit given in Fig. 2.21.

2.96 The waveforms given in Fig. 2.12 are applied to the circuit of Fig. 2.21. Deduce the output waveform F.

2.97 Show that
$$\overline{A}BC\overline{D} + \overline{A}BCD + A\overline{B}\overline{C}D + A\overline{B}CD + AB\overline{C}D + ABCD = AD + \overline{A}BC$$

2.98 Prove that $A \oplus B \oplus C = A \oplus C \oplus B$.

2.99 Express the function $F = A + BC$ in product-of-sum form.

2.100 Use a truth table to prove that
$$A + \overline{A}B = A + B$$

2.101 Use a truth table to prove that
$$A(\overline{A} + B) = AB$$

2.102 Evaluate $F = (A + B)(B + C)$ when
a) $A = B = C = 0$
b) $A = B = C = 1$
c) $A = 1, B = C = 0$.

2.103 Show that
a) $\overline{\overline{A}B} = A + \overline{A}B$
b) $\overline{A}\overline{B} + \overline{A}B = AB + \overline{A}\overline{B}$
c) $A(A + B) = A$

2.104 Simplify
a) $F = \overline{\overline{A + B} + \overline{A + B}}$
b) $F = \overline{\overline{A}\overline{B} + \overline{A + B}}$
c) $F = (AB + \overline{C})(A + BC)$
d) $F = (\overline{AC + D})(\overline{AC} + B)$

2.105 Show that $(AB + \overline{C}) + (\overline{A} + \overline{B})C = 1$.

2.106 Show that
$$B(\overline{AD + \overline{B}C})(\overline{A} + D)\overline{C}(\overline{A} + D) = 0.$$

2.107 Express $F = AB\overline{C} + A\overline{B}C + \overline{A}BC$ in product-of-sums form.

2.108 Express $F = (A + B + \overline{C})(A + \overline{B} + C)(\overline{A} + \overline{B} + \overline{C})$ in sum-of-products form.

2.109 Show that truth tables can be verified by the use of electronic lie detectors.

2.110 Show that
$$(B + D)(D + C)(A + D) = D + ABC$$

2.111 Show that
$$(B + D)(A + D)(C + B)(C + A) = AB + CD$$

2.112 Use a Karnaugh map to simplify
$$F = A\overline{B}C + \overline{A}B\overline{C} + \overline{A}BC + AB\overline{C} + ABC$$

2.113 Use a Karnaugh map to simplify
$$F = AB\overline{C} + ABC + \overline{A}B\overline{C} + \overline{A}\overline{B}\overline{C}$$
Then loop the squares marked with 0 to obtain the minimal solution for $\overline{F}$. Invert $\overline{F}$ check with the answer obtained in the first part of the question.

2.114 Use a Karnaugh map to minimize the equation
$$F = \overline{A}BC\overline{D} + \overline{A}B\overline{C}D + A\overline{B}CD + A\overline{B}C\overline{D} + ABCD + \overline{A}\overline{B}C\overline{D}$$

2.115 Use a Karnaugh map to minimize
$$F = (AB + \overline{B}\overline{C})(\overline{A}\overline{C} + \overline{A}BD + BCD)$$

2.116 Use a Karnaugh map to minimize
$$F = A\overline{B} + \overline{C}D + C + \overline{B}\overline{C}D$$

2.117 Use a Karnaugh map to minimize
$$F = \overline{A}\overline{B}C\overline{D} + \overline{A}\overline{B}CD + AB\overline{C}D + \overline{A}BCD + ABD + \overline{B}C\overline{D} + \overline{A}B\overline{C}D$$

2.118 Map the function
$$F = AB\overline{C} + A\overline{B}C\overline{D} + \overline{A}B\overline{C}D + ACD$$
Obtain the minimal solution for $\overline{F}$ by looping the 0 squares.

2.119 Map the function $F = AC + C\overline{D} + \overline{A}D + \overline{C}D$ and then simplify a) by looping 1's and b) by looping 0's.

2.120 Simplify a) algebraically and b) by mapping the function
$$F = (A + B)B(C + D) + (EB + AC)DB$$

2.121 Minimize $F = (\overline{\overline{A}B + AB})(\overline{A} + \overline{B})(A + B)$

2.122 Minimize $F = \overline{AB(B + C)} + \overline{AB(1 + C)}$

Use of NAND/NOR gates to generate AND/OR functions

Mark each of the following statements True or False.

2.123 The AND function can be generated using NAND gates but the OR function cannot.

2.124 The NOR gate with inverted inputs acts as an AND gate.

2.125 The NAND gate with inverted inputs acts as an OR gate.

2.126 The circuit shown in Fig 2.22 implements the AND function.

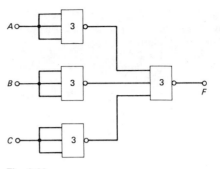

Fig. 2.23

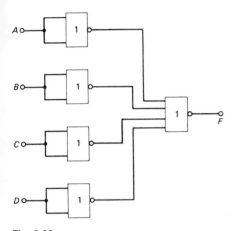

Fig. 2.22

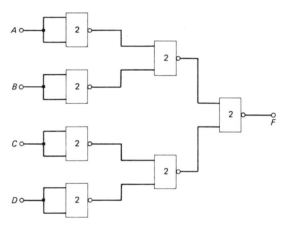

Fig. 2.24

2.127 Sum-of-product equations are easier to implement using NOR gates only than by using NAND gates only.

2.128 Implementation of the OR function using NOR gates requires a minimum of 3 gates.

2.129 Implementation of the AND function using NOR gates requires a minimum of 3 gates.

2.130 The use of *one* type of logic gate in a circuit is economically preferable.

2.131 A NAND gate is only employed when the NAND function is required.

2.132 The NOR function can be implemented using NAND gates only.

Multiple Choice Questions: mark the correct answer.

2.133 The NAND gate is often used to perform the AND logical function because *a)* it is the only type of gate readily available in i.c. form, *b)* the fan-in is larger, *c)* they are cheaper, faster and dissipate less power, or *d)* an unwarranted practice has now become common.

2.134 The AND function can be obtained by the cascade connection of *a)* two NOR gates, *b)* three NOR gates, *c)* two NAND gates, or *d)* three NAND gates.

2.135 The logic function performed by Fig. 2.23 is *a)* AND, *b)* OR, *c)* NAND, or *d)* NOR.

2.136 The logic function performed by Fig 2.24 is *a)* AND, *b)* OR, *c)* exclusive-OR, or *d)* none of these.

Short Exercises

2.137 Implement, using NAND gates only, the function $F = \overline{A + BC}$.

2.138 For your circuit in answer to **2.137** replace each NAND gate by a NOR gate and then determine the logical expression describing the new circuit.

2.139 Implement the circuit given in Fig. 2.25 using NOR gates only.

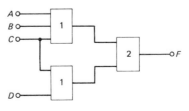

Fig. 2.25

2.140 For the circuit of Fig. 2.24 replace each NAND gate by a NOR gate and then determine the logical function performed by the new circuit.

2.141 Fig. 3.12 shows the pin connections of the 7400 quad 2-input NAND gate. Show the necessary connections to produce an OR function.

2.142 Implement the function $F = A + BC(A + C)$ using NAND gates only.

2.143 Implement the function $F = \overline{A}\overline{B} + BC + \overline{D}$ using *a*) NAND gates only and *b*) NOR gates only.

2.144 A circuit is required to produce the output $F_1 = AC + DB$ and $F_2 = \overline{A} + BD$. Draw a possible circuit using NAND gates only.

2.145 Implement using NAND gates only

$$F = (A + B + C)(\overline{A}\overline{C} + A\overline{B})$$

2.146 Implement using NOR gates only

$$F = ABC + \overline{A}\overline{B}C$$

Design of circuits from truth tables

Mark each of the following statements True or False.

2.147 A truth table shows the outputs obtained for all the possible combinations of the input variables.

2.148 The truth table of a 4-input gate should have four input combinations.

2.149 The truth Table 2.3 represents the expression

$$F = ABD + \overline{A}C\overline{D} + AB\overline{C}$$

Table 2.3

A	1 1 1 1 1 1 0 0 0 0 0 0 0 1
B	1 0 1 0 1 0 1 0 1 0 1 0 1 0
C	0 0 1 1 1 1 0 0 1 1 1 0 0 0
D	1 1 0 0 1 1 0 0 0 1 1 1 1 0
F	1 0 0 0 0 0 0 0 0 0 1 1 0 1

2.150 The Boolean expression for a function whose operation is described by a truth table is obtained by writing down each combination of input variables that gives $F = 1$ in the table.

Multiple Choice Questions: mark the correct answer.

2.151 The circuit whose truth table is given by Table 2.4 is shown in Fig. 2.26 *a*), *b*), *c*), or *d*).

Table 2.4

A	0 0 0 0 1 1 1 1
B	0 0 1 1 0 0 1 1
C	0 1 0 1 0 1 0 1
F	0 0 0 0 1 1 0 0

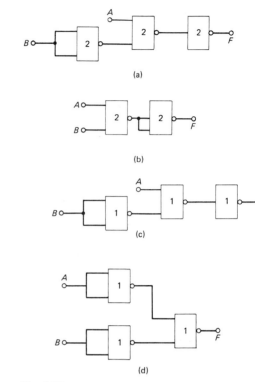

Fig. 2.26

2.152 The Boolean equation describing the operation of the circuit whose truth table is given by Table 2.5 is

a) $F = \overline{A}\overline{B}C + \overline{A}B\overline{C} + \overline{A}\overline{B}C + AB\overline{C}$

b) $\overline{ABC} + A\overline{B}C + \overline{A}B\overline{C} + AB\overline{C}$

c) $B + A + AB$, or *d*) $\overline{A + B + AB}$

Table 2.5

A	0 1 0 0 0 1 1 1
B	0 0 1 1 0 0 1 1
C	0 0 0 1 1 1 0 1
F	1 0 1 0 1 0 1 0

2.153 The Boolean expression representing the truth Table 2.6 is

Table 2.6

A	0 1 1 1 0 1 0 0
B	1 0 0 1 0 1 0 1
C	0 1 0 0 1 1 0 1
F	0 0 0 0 1 1 0 1

a) $F = A\overline{B}C + \overline{A}\overline{B}C + ABC$

b) $\overline{A}B\overline{C} + \overline{A}\overline{B}C + A\overline{B}C$

c) $\overline{A}BC + \overline{A}\overline{B}C + AB\overline{C}$

or *d*) $\overline{A}\overline{B}C + ABC + \overline{A}BC$

Short Exercises

2.154 A digital circuit has three inputs *A*, *B*, and *C* and three outputs *X*, *Y* and *Z*. The truth table for this circuit is given by Table 2.7. Determine the Boolean expressions for *X*, *Y* and *Z*.

Table 2.7

A	0 1 0 1 0 1 0 1
B	0 0 1 1 0 0 1 1
C	0 0 0 0 1 1 1 1
X	1 0 0 0 0 0 1 0
Y	0 1 1 0 1 1 0 0
Z	0 0 0 1 0 0 0 1

2.155 Obtain the Boolean equation for the output *F* of the circuit whose truth table is given by Table 2.8. Simplify the equation and then implement it using either NAND or NOR gates only.

Table 2.8

A	0 0 0 0 1 1 1 1
B	0 0 1 1 0 0 1 1
C	0 1 0 1 0 1 0 1
F	1 1 1 0 1 0 1 0

2.156 A circuit is required that has 10 inputs, labelled 0 through to 9, and one output. The output of the circuit should go high when or more of the inputs 1, 3 and 5 are high. Write down the truth table and hence determine the Boolean equation describing the wanted circuit. Implement the circuit using either NAND or NOR gates only.

2.157 Table 2.9 gives the truth table of a logic circuit.

Table 2.9

A	0 1 0 1 0 1 1 0
B	0 0 0 1 1 0 1 1
C	1 0 0 0 0 1 1 1
F	1 0 0 0 0 1 1 1

Obtain the Boolean equation describing the circuit and reduce it to its simplest form.

2.158 For the circuit in **2.157** write down the Boolean equation for $\overline{F}$. Then invert it to get *F*.

3 Practical electronic gates

Requirements of a logic gate

Mark each of the following statements True or False.

3.1 The speed of operation of a gate is the time that elapses before the output voltage changes in responses to an appropriate input condition.

3.2 The minimum power dissipation is always the most important consideration for a gate.

3.3 The propagation delay of a gate is always greater than its speed of operation.

3.4 When a rectangular pulse is applied to the base of a transistor its collector current does not immediately change.

3.5 The switching speed of a transistor can be increased by driving it into saturation.

3.6 The fan-in of a gate is the number of inputs that must be simultaneously high for the output to go high.

3.7 The noise immunity of a gate is always less than its noise margin.

3.8 The most popular version of the t.t.l. family is now the low-power Schottky.

3.9 As the fan-out of a gate is increased, so the logic 0 voltage level rises.

3.10 A quad 2-input NAND gate is quoted by the manufacturer as dissipating 2 mW per gate. The total power dissipated in the i.c. is 8 mW.

3.11 The fan-out of a gate is 5. This means that a minimum of 5 other gates must be connected to the output terminals.

3.12 The output waveform of a t.t.l. gate has a rise time of 0.1 ms. This is the time it takes the output voltage to rise from zero to +5 V.

3.13 When a transistor is fully ON it is said to be saturated.

3.14 The collector/emitter voltage of a saturated transistor is 0 V.

3.15 A Schottky diode is used to minimize the power dissipated in a circuit.

3.16 The main factor limiting the operating speed of a bipolar transistor is base charge storage but this factor does not affect c.m.o.s. devices.

3.17   The maximum output voltage of a gate for logic 0 is 2.0 V and the minimum voltage for logic 1 is 2.4 V. This means that the noise margin is 0.4 V.
3.18   The propagation delay of a gate increases with increase in the fan-out.

Multiple Choice Questions: mark the correct answer.
3.19   When a transistor is fully ON the collector/emitter voltage is labelled as *a*) the break-down voltage, *b*) the saturation voltage, *c*) the Schottky voltage, or *d*) the zener voltage.
3.20   A transistor has $V_{CE(SAT)} = 0.2$ V and it is used in a circuit with a collector supply voltage V_{CC} of +5 V and a collector load resistance of 1000 Ω. The saturated collector current is *a*) 5 mA, *b*) 0.2 mA, *c*) 20 mA, or *d*) 4.8 mA.
3.21   The switching speed of a transistor can be increased by the connection of a diode as shown by Fig. 3.1 *a*, *b*, *c*, or *d*.

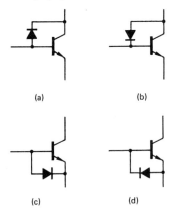

(a)   (b)

(c)   (d)

Fig. 3.1

3.22   A Schottky transistor *a*) dissipates less power, *b*) is cheaper, *c*) is faster, or *d*) is more readily available than an ordinary bipolar transistor.
3.23   A Schottky diode is a diode that has *a*) very low forward resistance, *b*) very high reverse resistance, *c*) high breakdown voltage, or *d*) very low charge storage.
3.24   The fan-out of a gate is *a*) the number of similar circuits it can drive, *b*) the number of output terminals it has, *c*) its maximum output voltage, or *d*) the number of its possible applications.
3.25   The mean current taken by a gate from the +5 V power supply is 0.5 mA. The power dissipated in the gate is *a*) 2.5 mW, *b*) 5 mW, *c*) 0.5 mW, or *d*) 0.
3.26   The noise margin of a gate is *a*) the r.m.s. noise at its output terminals, *b*) the input level that causes the output transistor to saturate, *c*) the maximum noise voltage that can appear at the input terminals without producing a change in the output state, or *d*) the maximum noise voltage that can appear at the input terminals without damaging the circuit.

3.27   A bipolar transistor has a collector supply voltage of 5 V and a collector resistance of 1200 ohms. When the transistor is saturated the collector current 4.0 mA. The saturation voltage of the transistor is *a*) 0.1 V, *b*) 0.2 V, *c*) 0.6 V, or *d*) 1 V.
3.28   The nominal supply voltage for gates in the t.t.l. family is 5 V ± 5%. The minimum and maximum voltages are *a*) 4.75 V and 5.25 V, *b*) 4.5 V and 5.5 V. *c*) 4.8 V and 5.2 V, or *d*) none of these.
3.29   The low-level noise margin of a gate is the difference between the maximum voltage that is recognized as *a*) 0 at the input and as 1 at the output, *b*) 1 at the input and 0 at the output, *c*) 0 at both the input and the output, or *d*) 1 at both the input and the output.

Short Exercises
3.30   The saturated collector current of a bipolar transistor of $h_{FE} = 60$ is 6 mA. When the transistor is turned full ON the base current is 120 μA. Calculate the excess base current and explain what happens to it.

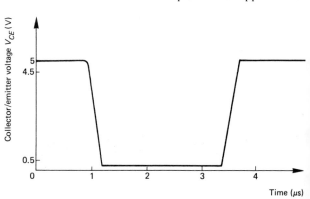

Fig. 3.2

3.31   Fig. 3.2 shows how the collector/emitter voltage of a transistor varies when a rectangular pulse of +5 V is applied to its base terminal. Calculate *a*) $V_{CE(sat)}$, *b*) the fall-time t_f, and *c*) the rise time t_r of the transistor.
3.32   When a rectangular pulse is applied to the base of a transistor Fig. 3.3*a*, the resultant change in the collector/emitter voltage is as shown in Fig. 3.3*b*. Give a name to both the time periods t_f and t_r and explain how they arise.
3.33   Fig 3.4 shows the output circuit of a gate. If the logical 1 level is defined by the voltage limits 3.5 V to 5 V calculate the fan-out of the circuit. Assume that each of the driven gates takes a current of 1.6 mA.
3.34   The gate shown in Fig. 3.4 is connected to four similar gates. Each gate has an input resistance of 4000 ohms. Calculate the logic 1 output voltage level.
3.35   A t.t.l. gate has a minimum logical 1 voltage level of 2.4 V and a maximum logical 0 voltage level of 400 mV. Calculate the worst-case noise margin of the gate.

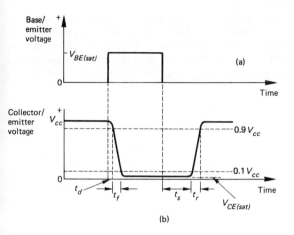

Fig. 3.3

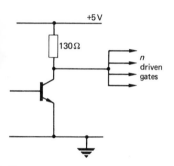

Fig. 3.4

3.36 A t.t.l. gate has a logical 1 voltage level of 3.0 V and a logical 0 voltage level of 0.4 V. Calculate *a*) the threshold voltage and *b*) the noise margin.

T.T.L., C.M.O.S. and E.C.L. families

Mark each of the following statements True or False.
3.37 The standard t.t.l. circuits employed a totem-pole output stage to increase both the speed of operation and the fan-out.
3.38 The Schottky NAND and NOR gates use a totem-pole output stage to increase both the operating speed and the fan-out.
3.39 The fan-out of a t.t.l. gate is 25.
3.40 The fan-out of a c.m.o.s. gate may be in the range 20 to 50.
3.41 The noise margin of an open-collector gate is better than that of a totem-pole output gate.
3.42 The LS 7402 is an example of a low-power Schottky gate.

3.43 Both standard and low-power Schottly gates employ a multiple-emitter transistor input.
3.44 The wired-OR connection can be used with open-collector NOR gates.
3.45 C.M.O.S. gates use both n-channel and p-channel enhancement-mode mosfets.
3.46 The main advantage claimed for the c.m.o.s. logic family is a low power dissipation.
3.47 The output of a t.t.l. gate can be directly connected to the input of a c.m.o.s. gate.
3.48 The output of a c.m.o.s. gate can be directly connected to the input of a t.t.l. gate.
3.49 Devices in the e.c.l. logic family are used when the most important factor is the need for minimum power dissipation.
3.50 An e.c.l. gate can be directly connected to a t.t.l. gate but not a c.m.o.s. gate.
3.51 E.C.L. is faster than both t.t.l. and c.m.o.s. and it is therefore used when the fastest possible speed is wanted.
3.52 When a c.m.o.s. gate is in either its high or its low output state it will dissipate zero power.
3.53 The 7400 is a quad 2-input NAND gate in the t.t.l. family.
3.54 A pull-up resistor is used when the current taken from the power supply must be limited.
3.55 Gates in different sub-divisions of the t.t.l. family can be directly interfaced to one another.
3.56 When a large current output is required, two totem-pole stages can have their outputs connected in parallel.

Multiple Choice Questions: mark the correct answer.
3.57 The fastest logic family is *a*) standard t.t.l., *b*) low-power Schottky t.t.l., *c*) c.m.o.s., or *d*) e.c.l.
3.58 The logic family that dissipates the least power is *a*) standard t.t.l., *b*) low-power Schottky t.t.l., *c*) c.m.o.s., or *d*) e.c.l.
3.59 If the number of gates connected to the output terminal of a gate is greater than its fan-out, then *a*) both the logic 1 and logic 0 voltage levels will be too high, *b*) both the 1 and 0 voltage levels will be too low, *c*) the 1 voltage level will be too high and the 0 level too low, or *d*) the 1 voltage level will be too low and the 0 voltage level will be too high.
3.60 A 2-input NAND gate needs three i.c. pins. How many such gates can be accommodated in a 14-pin d.i.l. package: *a*) 4, *b*) 3, *c*) 2, or *d*) 1?
3.61 The very high switching speed of gates in the e.c.l. logic family is obtained by *a*) the use of Schottky transistors, *b*) the use of Schottky clamping diodes, *c*) the use of negative logic, or *d*) ensuring that transistor saturation does not occur.
3.62 In the e.c.l. family the logic levels 1 and 0 are represented by *a*) +5 V and −5 V, *b*) −5 V and 0 V, *c*) −0.9 V and −1.75 V, or *d*) 0.9 V and 1.75 V.

3.63 The speed of a c.m.o.s. gate is lower than that of the other families. This is because *a*) the input current is small, *b*) the d.c. power supply voltage is not fixed at a standard value, *c*) the combination of high input impedance and stray capacitances gives a relatively long time constant, or *d*) the mosfets easily saturate.

3.64 The fan-in of the circuit shown in Fig. 3.5 is *a*) 4, *b*) 2^4, *c*) 1, or *d*) 4^2.

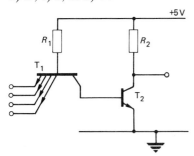

Fig. 3.5

3.65 When all the inputs to the circuit of Fig. 3.5 are at $+5\,V$ the input transistor T_1 has *a*) both its collector/base and emitter/base junctions reverse biased, *b*) its collector/base junction is forward biased but the emitter/base junction is reverse biased, *c*) both these junction are forward biased, or *d*) the collector/base junction is reverse biased but the emitter/base junction is forward biased.

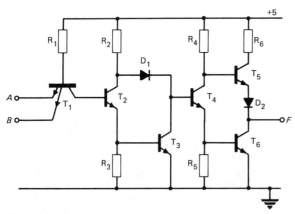

Fig. 3.6

3.66 The circuit given in Fig. 3.6 is *a*) a NAND, *b*) a NOR, *c*) an AND, or *d*) an AND-OR-INVERT gate.

3.67 The expression $F = \overline{ABCD + EFGH}$ describes an AOI gate that is *a*) 2-input 4-wide, *b*) 4-input 2-wide, *c*) 2-input 2-wide, or *d*) 4-input 2-wide.

3.68 The logical function performed by two open-collector NAND gates with their outputs paralleled is

a) $\overline{ABCD}$, *b*) $\overline{A + B + C + D}$
c) $\overline{AB + CD}$, or *d*) $\overline{AB(C + D)}$

3.69 AOI gates are described as *n*-wide, *m*-input. The term "wide" denotes *a*) the number of inputs to the circuit, *b*) the number of inputs to the NOR gate, *c*) the physical width of the i.c. package, or *d*) the number of inputs that must go high for the output to go high.

Short Exercises

3.70 Draw the block diagram of a 2-wide 2-input AOI gate.

3.71 Fig. 3.7 shows the circuit of a c.m.o.s. gate. State the logical function performed by the circuit and describe its operation.

3.72 Fig. 3.8 shows the pin connections of the 4012 dual 4-input NAND gate. Show how the i.c. would be connected to produce the logical function $F = (A + B + C)D$.

3.73 Explain why c.m.o.s. devices require very careful handling whilst being fitted into a circuit. List the precautions you would take.

3.74 What is likely to happen if one input is left disconnected on *a*) a standard t.t.l. gate, *b*) a low-power Schottky gate, *c*) a c.m.o.s. gate, or *d*) an e.c.l. gate?

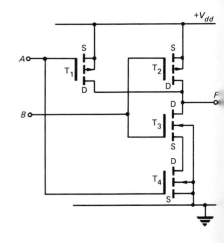

Fig. 3.7

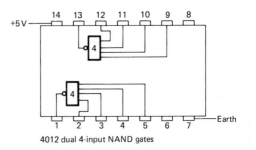

4012 dual 4-input NAND gates

Fig. 3.8

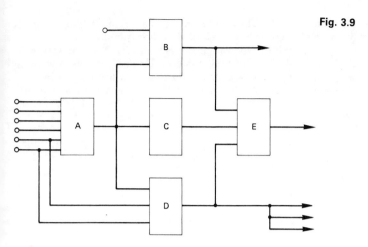

Fig. 3.9

3.75 Refering to Fig. 3.9. What are *a*) the required fan-in for each gate, *b*) the required fan-out for each gate?

3.76 Explain why power dissipation in a digital i.c. is of importance. If c.m.o.s. is ruled out in the grounds of speed what is then the best family in this respect?

3.77 Explain the significance of noise margin for a gate and say how it is related to the logic levels for binary 1 and 0.

3.78 Explain why a disconnected input to a t.t.l. gate acts like a logical 1 input.

3.79 A 24-input NAND gate is to be designed using 7430 8-input NAND gate i.c.s. Draw a suitable circuit. If necessary, 7404 hex inverters and 7420 dual 4-input NAND gates are also available.

3.80 Explain the reasons why the AND gate, although available in the t.t.l. and c.m.o.s. logic families, is rarely used and any wanted AND functions are produced by suitably connected $\overline{NAND}$ or NOR gates.

3.81 The logical function $F = \overline{ABC + DEF}$ is to be implemented. Draw possible arrangements using *a*) NAND gates only, *b*) open-collector gates only.

4 Bistable multivibrators

Mark each of the following statements True or False.

4.1 An S-R flip-flop can be connected to act as a D flip-flop.

4.2 A J-K flip-flop cannot be connected to act as T flip-flop.

4.3 The J-K flip-flop can be connected to act as a divide-by-2 circuit.

4.4 It is not possible to convert an S-R flip-flop into a J-K flip-flop.

4.5 An S-R flip-flop cannot be converted into a T flip-flop.

4.6 Fig. 4.1 will act as a divide-by-2 circuit.

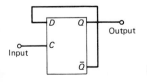

Fig. 4.1

4.7 Table 4.1 is the truth table of a D flip-flop.

Table 4.1

D	0	0	1	1
Q	0	1	0	1
Q^+	0	1	1	0

4.8 Only J-K flip-flops can be purchased in i.c. packages.

4.9 When a set D flip-flop is cleared, the Q terminal becomes logical 1.

4.10 A J-K flip-flop is in the reset state and is then cleared. The $\bar{Q}$ terminal is then a logical 0.

4.11 In a master-slave J-K flip-flop the function of the slave is to operate the master.

4.12 Some i.c. flip-flops are provided with a pre-set pin. This allows the flip-flop to be set independently of the conditions of the *J* and *K* inputs and the clock.

4.13 Some i.c. D flip-flop have a clear pin. This enables the circuit to be set independently of the conditions on the D and the clock inputs.

4.14 The operation of a J-K flip-flop is indeterminate when $J = K = 1$.

4.15 The operation of an S flip-flop is indeterminate when $S = R = 1$.

4.16 A J-K flip-flop toggles on each clock pulse if its J and K inputs are both connected to logical 0.

4.17 A quad 2-input NOR gate cannot be connected to operate as a divide-by-2 circuit.

4.18 The S-R flip-flop is said to be a sequential circuit because its outputs are not solely determined by its imputs.

4.19 The leading edge of a clock pulse is when the pulse changes from 0 to 1.

4.20 A leading-edge triggered flip-flop operates when the clock changes from 1 to 0.

4.21 Referring to Fig. 4.2, the Q output will go to logical 0 when the leading edge of the next clock pulse occurs.

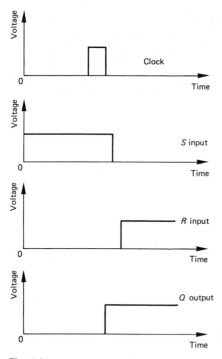

Fig. 4.2

4.22 When the J and K inputs of a J-K flip-flop are both held at logical 1 the circuit will toggle at each clock pulse.

4.23 At the end of the K pulse in Fig. 4.3 the Q output will be at logical 0.

4.24 Master-slave flip-flop are edge-triggered devices.

4.25 The J and K inputs of a J-K flip-flop are normally kept constant whilst the clock is high.

4.26 When a flip-flop is preset its Q output goes high.

4.27 C.M.O.S. flip-flops do not generally use cross-coupled NAND or NOR gates to form a flip-flop.

4.28 A J-K flip-flop is described by the manufacturer as having a toggle frequency of 150 MHz. This is the maximum allowable clock frequency.

4.29 The function table of a flip-flop (74LS76) is shown by Table 4.2. When the preset or the clear inputs are low they reset the circuit whatever the values of the other inputs.

Table 4.2

Preset	Clear	Clock	J	K	Q	$\bar{Q}$
L	H	X	X	X	H	L
H	L	X	X	X	L	H
L	L	X	X	X	H	H

Multiple Choice Questions: mark the correct answer.

4.30 A J-K flip-flop has its J and K inputs connected to the logic 1 voltage level. The circuit toggles at the leading edge of each clock pulse. The flip-flop is *a*) a leading-edge triggered type, *b*) a trailing-edge triggered type, *c*) a master-slave type, or *d*) faulty.

4.31 When an S-R flip-flop is connected as a D flip-flop the race condition is not possible because *a*) the switching speed is reduced, *b*) the switching speed is increased, *c*) the noise margin is reduced, or *d*) the indeterminate state $S = R = 1$ cannot occur.

4.32 The $\bar{Q}$ output of a J-K flip-flop follows the sequence 0,1,1,0,0,1. The J-K inputs must have followed the sequence
   a) 11,11,11,11,11, *b*) 01,01,11,10,11
   c) 10,01,10,11,01, or *d*) 10,10,01,01,11.
Assume the flip-flop is initially set.

4.33 When a flip-flop is said to toggle it *a*) has its Q output change state at each clock pulse, *b*) has its Q output remain constant at logical 1, *c*) has its Q output remain constant at logical 0, or *d*) operates non-synchronously.

4.34 The waveforms shown in Fig. 4.2 relate to *a*) a basic S-R flip-flop, *b*) a leading-edge triggered S-R flip-flop, *c*) a trailing-edge triggered S-R flip-flop, or *d*) a master-slave flip-flop.

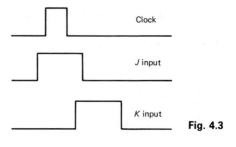

Fig. 4.3

4.35   An S-R flip-flop is initially cleared. The sequence of input S and R pulses is 00,01,01,10,01. The Q output follows the sequence
   a) 0,1,1,0,1, b) 0,0,0,1,0
   c) 1,0,1,0,1, or d) 1,1,1,0,1.

4.36   The disadvantage of the S-R flip-flop is its indeterminate operation when $S = R = 1$. This means that a) both the Q and $\bar{Q}$ outputs may become 1 at the same time, b) both the Q and the $\bar{Q}$ outputs may become 0 at the same time, c) the circuit will be unable to change state, or d) the next state of the circuit cannot be predicted.

4.37   An S-R flip-flop is in its reset state. The input state $S = 1$, $R = 0$ is then applied and then the circuit is cleared. If the input state $S = 0$, $R = 1$ is now applied the flip-flop will a) change state to $Q = 1$, $\bar{Q} = 0$, b) change state to $Q = 0$, $\bar{Q} = 1$, c) remain in the state $Q = 1$, $\bar{Q} = 0$, or d) remain in the state $Q = 0$, $\bar{Q} = 1$.

4.38   One of the circuits shown in Fig. 4.4 acts as a divide-by-2 circuit. Is it a, b, c, or d?

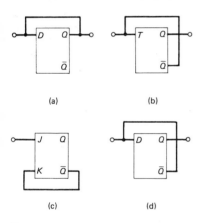

(a)   (b)

(c)   (d)

Fig. 4.4

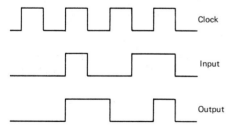

Clock

Input

Output

Fig. 4.5

4.39   The waveforms shown in Fig. 4.5 relate to a) an S-R flip-flop, b) a J-K flip-flop, c) a D flip-flop, or d) a T flip-flop.

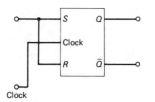

Clock

Fig. 4.6

4.40   The circuit shown in Fig. 4.6 will act as a) a D flip-flop, b) a T flip-flop, c) both a D and a T flip-flop, or d) neither a D nor a T flip-flop.

4.41   Table 4.3 is the truth table of a) a D flip-flop, b) a T flip-flop, c) an S-R flip-flop, or d) a J-K flip-flop.

Table 4.3

0	0	0
0	1	0
1	0	1
·1	1	1

4.42   A D flip-flop can be made from an S-R flip-flop by a) connecting its S and R terminals together and applying the input to the common terminals, b) connecting an inverter between the S and R terminals and applying the input to the S terminal, c) connecting an inverter between the Q and $\bar{Q}$ terminals and applying the input to the R terminal, or d) connecting the $\bar{Q}$ terminal to the R terminal and applying the input to the S terminal.

4.43   Master-slave operation of a J-K flip-flop is used to a) ensure that the circuit switches at the correct times, b) increase the speed of switching, c) increase the power dissipation, or d) reduce the noise margin.

Short Exercises

4.44   Give the circuit symbol for a clocked J-K flip-flop and write down its truth table.

4.45   Draw the circuit of a NAND gate S-R flip-flop and explain its operation.

4.46   A J-K flip-flop is in the state $Q = 0$, $\bar{Q} = 1$. What change in state will occur if the flip-flop is a) set, b) reset, or c) cleared?

4.47   Write down the truth table of a J-K flip-flop. Now let $J = K$ and then determine the corresponding Q values. What kind of circuit does the truth table now represent?

4.48   Fig. 4.7 shows an S-R flip-flop whose S and R inputs are fed from two AND gates. One of the two inputs to each gate is obtained from the output of the flip-flop as shown. Write down the truth table of the circuit and hence determine what kind of circuit it is.

4.49   Draw the block diagram of a master-slave J-K flip-flop and explain its operation. Point out how the race condition to which Fig. 4.7 is prone is eliminated.

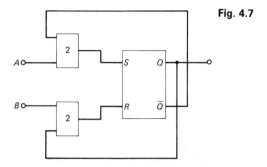

Fig. 4.7

4.50 The waveform shown in Fig. 4.8 is connected to the input terminal of Fig. 4.4*d*. Draw the output waveform.

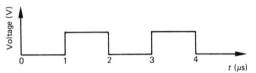

Fig. 4.8

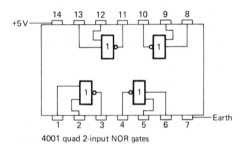

4001 quad 2-input NOR gates

Fig. 4.9

4.51 Fig. 4.9 shows the pin connections of the 4001 quad 2-input NOR gate i.c. Show how it can be connected to act as a D flip-flop.
4.52 Discuss the reasons why some S-R flip-flops are clocked.
4.53 Why are most J-K flip-flops either master-slave or edge-triggered types?
4.54 What are the functions of *a*) the clear and *b*) the reset terminals when provided on a flip-flop? A particular device has clear and reset terminals that are active low. If $J = 1$, $K = 0$ and the clear is at 0, what is the state of the Q output when reset = 1?
4.55 Fig. 4.10 shows the input and output waveforms of a flip-flop. Draw either a D or a J-K flip-flop connected to give this result.
4.56 The pin connections of the 7474 dual D type positive-edge triggered flip-flop are shown in Fig. 4.11. Identify the function of each pin.
4.57 The function table of the 7474 D flip-flop is given in Table 4.4. *a*) If the Preset input is low, does the

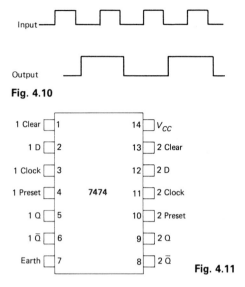

Fig. 4.10

Fig. 4.11

flip-flop set whatever the state of the D and the clock inputs? *b*) If the Preset and Clear inputs are not used, should they be connected to +5 V or to 0 V? *c*) If the D input is high, and the Preset and Clear are not used, does the flip-flop set or reset when the clock changes from 0 to 1?

Table 4.4

Preset	Clear	Clock	D	Q^t	$\bar{Q}^t$
L	H	X	X	H	L
H	L	X	X	L	H
L	L	X	X	H	H
H	H	0–1	H	H	L
H	H	0–1	L	L	H
H	H	L	X	Q	$\bar{Q}$

4.58 Draw the circuit of a gated S-R flip-flop using NAND gates. If the S and R input waveforms are as shown in Fig. 4.2 draw the output waveform if the circuit is arranged to be *a*) leading-edge triggered and *b*) trailing-edge triggered.
4.59 Fig. 4.12 shows the input waveforms of a J-K flip-flop. Draw the output waveforms if the circuit is *a*) leading-edge triggered and *b*) trailing-edge triggered.

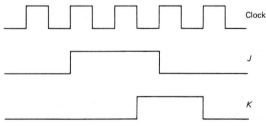

Fig. 4.12

5 Counters

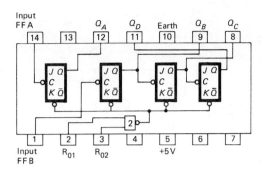

Fig. 5.4

Mark each of the following statements True or False.

5.1 A J-K flip-flop cannot be used as a divide-by-2 circuit.

5.2 Counters may be either synchronous or non-synchronous types.

5.3 Non-synchronous counters are faster to operate than synchronous counters because they do not have to wait for each clock pulse to arrive.

5.4 Fig. 5.1 shows the circuit of a divide-by-16 ripple counter.

5.5 For the circuit given in Fig. 5.2 to operate as a divide-by-8 circuit, each of the J and K inputs must be connected to the logical 1 voltage level.

5.6 If, in Fig. 5.2, flip-flops A and C are set and flip-flop B is reset, the count is 2.

5.7 If, in Fig. 5.2 the clear line goes high, all three flip-flops will be reset.

5.8 The circuit shown in Fig. 5.3 is a 4-bit synchronous counter.

5.9 The count of the circuit given in Fig. 5.3 can be reduced to 10 by taking the outputs of some stages via a NAND gate to the common clear line.

5.10 A decade counter counts from 0 to 100 in steps of 10.

5.11 All counters are triggered on the trailing edges of the clock pulses.

5.12 Fig. 5.4 shows the pin connections of the 7493 counter. The i.c. cannot be connected to operate as a divide-by-16 counter.

5.13 Decoding the outputs of each stage of a counter allows each count to give a unique output.

5.14 A dynamic hazard may occur at the decoded output of a counter if the individual flip-flops do not change state simultaneously.

5.15 Glitch is another name for dynamic hazard.

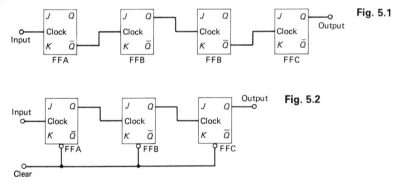

Fig. 5.1

Fig. 5.2

Fig. 5.3

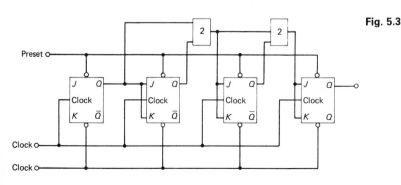

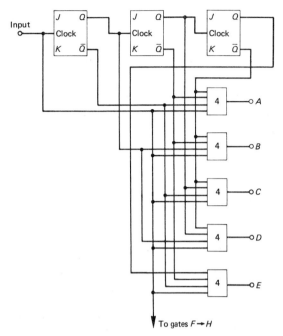

Fig. 5.5

Fig. 5.5 shows a 3-bit counter with decoded outputs.

5.16 The circuit is a synchronous counter.

5.17 The decoded output A is high when the count is 0.

5.18 The decoded output B is high when the count is 3.

5.19 The decoding gates have a clock input to assist in the production of glitches.

5.20 If a decoded count of 7 is required and AND gate with inputs $Q_A\bar{Q}_BQ_C$ and clock is needed.

5.21 A Schmitt trigger can be used to convert the output of a crystal oscillator to a rectangular clock waveform.

5.22 A synchronous counter cannot have its count reduced below 2^n, where $n =$ number of stages.

5.23 A ripple counter cannot be produced using T flip-flops.

5.24 The circuit shown in Fig. 5.5 could be used as a binary-to-decimal converter.

5.25 A ripple counter consists of six J-K cascaded flip-flops. Its maximum count is 63.

5.26 In a presettable counter the count may start at any number other than 0.

5.27 In a synchronous counter a particular stage toggles on a clock pulse only if all the less significant stages are at $Q = 1$.

5.28 For a D flip-flop counter the Q output of each stage is the same as the D input at the previous clock pulse.

5.29 An up-down counter is one in which the count may go up or down depending upon the speed of the clock.

5.30 Up-down counters are available in the c.m.o.s. and t.t.l. families.

5.31 In a ripple counter a stage will only change state if the previous or less significant stage changes state.

5.32 A decoder can be used in conjunction with a counter to determine when the maximum count has been reached.

5.33 The count of some synchronous i.c. counters can be reduced to less than 2^n by modifying the count length with the preset inputs.

5.34 The count of other synchronous i.c. counters can be modified by clearing the circuit when the desired maximum count is reached.

Multiple Choice Questions: mark the correct answer.

5.35 The circuit shown in Fig. 5.6 is *a*) a 3-bit non-synchronous counter, *b*) a 3-bit synchronous counter, *c*) a 3-bit synchronous up-down counter, or *d*) a shift register.

5.36 The maximum count of the circuit shown in Fig. 5.7 is *a*) 8, *b*) 10, *c*) 15, or *d*) 16.

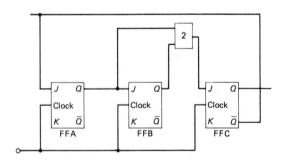

Fig. 5.6

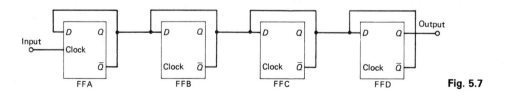

Fig. 5.7

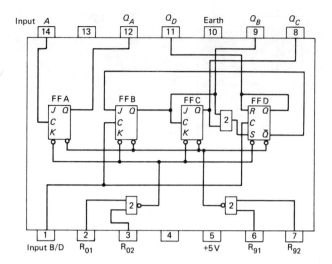

Fig. 5.8

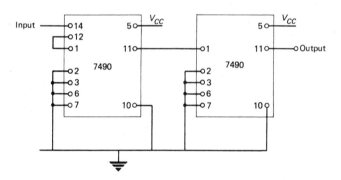

Fig. 5.9

5.37 The maximum count of a divide-by-7 counter is *a*) 7, *b*) 8, *c*) 4, or *d*) 6.

5.38 A 5-stage counter can give a maximum count of *a*) 5, *b*) 10, *c*) 32, or *d*) 31.

5.39 Fig. 5.8 shows the pin connections of the 7490 counter. If pin 11 is connected to pin 14 and pin 1 is used as the input terminal, the output at pin 12 gives *a*) divide-by-10, *b*) divide-by-8, *c*) divide-by-5, or *d*) divide-by-9 counter.

5.40 The 4-bit counter of Fig. 5.1 has the Q_B and Q_D outputs connected to the inputs of a 2-input AND gate. The output of the AND gate is connected to the active high clear terminals of the four flip-flops. The circuit acts as a divide-by- *a*) 10, *b*) 8, *c*) 19, or *d*) 12 counter.

5.41 A NAND Schmitt trigger consists of *a*) a Schmitt

trigger followed by a NAND gate, *b*) a Schmitt trigger preceded by a NAND gate, *c*) a Schmitt trigger that can also act as a NAND gate, or *d*) a NAND gate that can also be connected to act as a Schmitter trigger.

5.42 A semi-synchronous counter has *a*) only the first stage clocked, *b*) all stages except the final stage clocked, *c*) only the final stage clocked, or *d*) the clock pulse applied only when the maximum count is reached.

5.43 The circuit shown in Fig. 5.9 is a divide-by- *a*) 10, *b*) 20, *c*) 100, or *d*) 50 counter.

5.44 In a synchronous counter *a*) the stages operate one after the other, *b*) the maximum count is 16, *c*) all stages operate at the same time, or *d*) a clock is not needed.

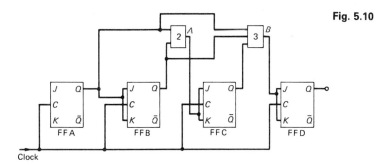

Fig. 5.10

5.45 The circuit shown in Fig. 5.10 is *a*) ÷16 ripple counter, *b*) ÷16 synchronous counter, *c*) ÷12 synchronous counter, or *d*) none of these.

Short Exercises
5.46 Show how one 7490 and one 7493 i.c. counter can be connected together to give a count of 80. The pin connections of these devices are given in Figs. 5.8 and 5.4.
5.47 Two 7490 decade counters are to be connected to give a count of 21. Obtain the required circuit.
5.48 Three decade counters are connected together to give a count of 1000. Calculate the frequency of the output waveform when the clock frequency is 3 MHz. What is the most likely source of the clock waveform?
5.49 Write down the truth table for the circuit shown in Fig. 5.7 and hence determine its count.
5.50 The 7424 hex Schmitt trigger i.c. has a positive-going threshold voltage of 1.71 V and a negative-going threshold voltage of 0.88 V. Calculate *a*) its hysteresis and *b*) the duration of the output pulse when a 2 V 500 kHz sinusoidal signal is applied to the input terminals.

5.51 Determine the count of the circuit given in Fig. 5.11.
5.52 Determine the count of Fig. 5.12.
5.53 Show how a 4-bit synchronous counter can have its count reduced to 9 by using the reset terminal of each flip-flop.
5.54 The c.m.o.s. 4020 counter has 14 stages. What is its maximum count? Why does it have only 12 Q outputs?
5.55 Sketch the Q_A, Q_B, Q_C and Q_D waveforms of the counter of Fig. 5.12.
5.56 Write down the truth table of a 4-bit synchronous counter and use it to show the need for two AND gates.
5.57 The counter of **5.55** is to be extended to have 5 stages. Sketch the extra circuitry required.
5.58 Write the truth table of the circuit given in Fig. 5.7.
5.59 Draw the clock and output waveforms of a divide-by-8 counter.
5.60 Discuss the different ways in which the count of a 4-bit non-synchronous counter can have its count reduced.

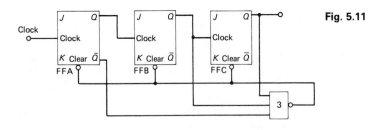

Fig. 5.11

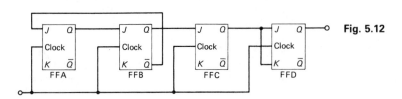

Fig. 5.12

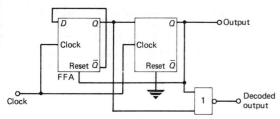

Fig. 5.13   **Fig. 5.14**

5.61 Show how a number of 7493 counters can be connected to give a count of 50.

5.62 Three J-K flip-flops are connected to operate as a ripple counter. Give the counting sequence obtained at *a*) the Q outputs and *b*) the $\bar{Q}$ outputs.

5.63 List the relative merits of t.t.l. and c.m.o.s. counters.

5.64 Determine the *a*) the maximum count and *b*) the decoded count of the circuit given in Fig. 5.13.

5.65 Determine the count of the circuit of Fig. 5.14.

5.66 The c.m.o.s. 4516 is described as a presettable up-down binary counter. Explain what this means.

5.67 Several i.c. counters are said by the manufacturer to be synchronous counters with asynchronous clear. What does this mean?

5.68 A 5-stage ripple counter has an input clock frequency of 1 MHz. At what frequency do the second, third and fifth stages operate?

5.69 Write down the truth table for the circuit given in Fig. 5.7. Assume forward-edge triggering.

6 Memories

Mark each of the following statements True or False.

6.1 A volatile memory has the ability to retain stored data when the power supply is switched off.

6.2 A 1 kilobit memory contains 1000 locations.

6.3 A 64 kilobit memory contains 65,000 locations.

6.4 The basic requirements for a ram are that *a*) any location can be addressed and *b*) data can be read out of, or written into, an addressed location.

6.5 Registers are used to provide short-term memory.

6.6 A cassette tape is an example of a read-only memory or rom.

6.7 Fig. 6.1 shows a shift register.

6.8 Dynamic rams use a large number of fast-operating flip-flops as memory cells.

6.9 All roms are non-volatile.

6.10 A prom is a programmable read-only memory.

6.11 Once a prom has been programmed it cannot be altered.

6.12 Once an eprom has been programmed it cannot be altered.

6.13 In an eprom the logical 1 state is stored at a location by the storage of an electrical charge.

6.14 In a prom the logical 0 state is stored at a location by the blowing of a fuse.

6.15 All proms and eproms are members of one branch or the other of the t.t.l. family.

6.16 A ram is used for the permanent storage of data.

6.17 Shift registers are operated with either serial- or parallel-inputted data.

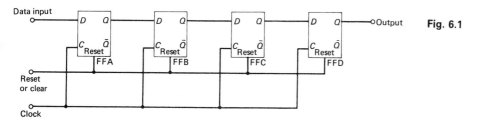

Fig. 6.1

6.18 All shift registers can accept either serial or parallel input data.

6.19 Bipolar rams have longer access times than mosfet rams.

6.20 In a dynamic ram, data is stored as long as the power supply is switched on.

6.21 A static ram must be periodically refreshed to retain its stored data.

6.22 A rom is organized as 4096×8. It will have 12 address lines.

6.23 The rom in **6.22** has 3 data lines.

6.24 Serial loading of a shift register means that data is entered one bit per clock pulse.

6.25 A rom has 256 memory locations. The addresses range from 0 to 256.

6.26 A required memory location in a rom or a ram is selected by means of a single decoder.

6.27 A number of 1024×1 rams are to be interconnected to produce a 1024×8 ram, so that 8 rams are needed.

6.28 Both ram and rom are random access memories.

6.29 A rom is said to organized as 256×4. It is a 1 kilobit memory.

6.30 The rom in **6.29** can store 4-bit words.

6.31 The rom in **6.29** has 8 address pins.

6.32 The rom in **6.29** has 3 data pins.

6.33 The rom in **6.29** requires an i.c. package with 11 pins.

6.34 A 65 kilobit ram can store 65 536 bits of data.

6.35 A rom is programmed by the manufacturer to perform some particular function and this cannot be altered.

6.36 For the ram shown in Fig. 6.2 the number of addressable locations is 16.

6.37 For the ram of Fig. 6.2 the number of data in/out pins is 4.

6.38 Referring to Fig. 6.3 the two $\bar{s}$ pins select the chip when low.

6.39 The ram of Fig. 6.3 is organized as 256×4.

6.40 Some shift registers can have an 8-bit word loaded simultaneously and then transmitted to an external circuit one bit at a time.

6.41 The arrangement shown in Fig. 6.4 could be used to transmit a 4-bit word over a single line.

6.42 Both ram and rom are available in both static and dynamic form.

6.43 Dynamic memories are available with densities about four times greater than static memories.

6.44 Dynamic ram is often known as dram.

6.45 All ram chips have a chip-select pin which, when active, enables the memory.

6.46 In a shift register the least significant stage is normally the right-hand one.

6.47 A shift register cannot be used as a serial-to-parallel converter.

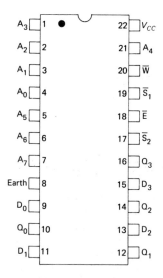

Fig. 6.3

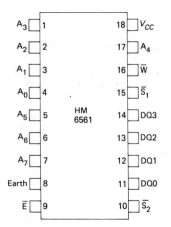

Fig. 6.2

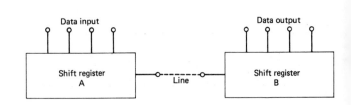

Fig. 6.4

Multiple Choice Questions: mark the correct answer.

6.48 The main difference between a dynamic shift register and a static shift register is that *a*) static types are only available in t.t.l. technology, *b*) dynamic types are slower, *c*) dynamic types are of smaller capacity, or *d*) dynamic types need periodic refreshing.

6.49 The i.c. shown in Fig. 6.5 is *a*) a ram, *b*) a rom, *c*) a shift register, or *d*) a 4-bit counter.

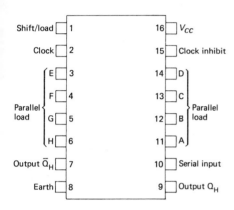

Fig. 6.5

6.50 An address decoder has 3 input lines. The number of output lines is *a*) 3, *b*) 2, *c*) 8, or *d*) 16.

6.51 A ram i.c. always needs one input additional to the requirement of a rom i.c. This is *a*) chip select $\overline{CS}$, *b*) address line A, *c*) two positive V_{cc}, or *d*) read/write.

6.52 One of the advantages of bipolar static rams is *a*) high speed, *b*) high power dissipation, *c*) maximum packing density, or *d*) very low supply voltage requirement.

6.53 C.M.O.S. memory chips are *a*) very fast, *b*) not available, *c*) very expensive, or *d*) of very low power dissipation.

6.54 The organization of a ram is *a*) the transistor technology employed, *b*) the number of package pins, *c*) the product number of bits × number of words, or *d*) the product access time × number of locations.

6.55 A 1024 × 1 ram has *a*) 1, *b*) 1024, *c*) 10, or *d*) 6 address lines.

6.56 A number of 1024 × 1 rams are to be interconnected to product a 8192 × 4 ram. The number of rams is *a*) 4, *b*) 8, *c*) 2, or *d*) 32.

6.57 In the circuit of Fig. 6.4 the shift registers A and B must be *a*) both PISO, *b*) both SIPO, *c*) A is PISO and B is SIPO, or *d*) A is SIPO and B is SISO.

6.58 The number of memory locations that 12 address lines can access is *a*) 1024, *b*) 2048, *c*) 4096, or *d*) 12.

6.59 A 2 kilobit rom is organized as 256 8-bit words. It has *a*) 256 address lines and 8 data lines, *b*) 8 address lines and 8 data lines, *c*) 9 address lines and 3 data lines, or *d*) 7 address lines and 4 data lines.

6.60 The highest address in a rom is 1023. The rom contains *a*) 1023, *b*) 1024, *c*) 1000, or *d*) 23 locations.

6.61 A ram contains 16 384 locations organized as 2048 8-bit words. The ram must have *a*) 10 address lines and 8 data lines, *b*) 1024 address lines and 8 data lines, *c*) 10 address lines and 3 data lines, or *d*) 11 address lines and 8 data lines.

6.62 An eprom is *a*) a volatile rom, *b*) a ram whose programming is permanent, *c*) a rom whose programming can be changed, or *d*) an experimental device.

6.63 Semiconductor memories have replaced magnetic core memories because *a*) suitable cores are no longer available, *b*) core memories are more expensive, bulkier, and slower, *c*) core memories cannot interface with t.t.l. circuitry, or *d*) core memories are volatile.

6.64 Dynamic ram has the advantage over static ram in that it *a*) requires periodic refreshing, *b*) occupies a greater chip area, *c*) is not subject to noise, or *d*) it is cheaper and faster to operate.

6.65 A memory is said to be volatile if it *a*) loses its data when the power supply is removed, *b*) is easily affected by external noise sources, *c*) has a short access time, or *d*) is affected by external magnetic fields.

6.66 The letters SIPO applied to a shift register indicate *a*) serial-in/parallel-out, *b*) signal-input/power-output, *c*) set-input/preset-output, or *d*) Schottky-input/pnp-output.

6.67 The circuit shown in Fig. 6.1 is *a*) a ripple counter, *b*) a shift register, *c*) a synchronous counter, or *d*) a ram.

6.68 The access time of a memory is *a*) the time it takes to operate after the power supplies are switched on, *b*) the time it takes for a maintenance man to gain access, *c*) the time taken for a particular location to be addressed, or *d*) the time taken for one word to be read out of the memory.

6.69 A ram *a*) has the same access time for all locations, *b*) has a shorter access time for low-numbered locations than for high-numbered locations, *c*) can only be manufactured using magnetic elements, or *d*) is usually too expensive to be used.

6.70 A location in a rom *a*) can only be read, *b*) can only be written, *c*) can be both read out of and written into, or *d*) can neither be read out of nor written into.

6.71 Fig. 6.6 shows a diode rom. When the input address is 11 the output is *a*) $\overline{D}_0 D_1 D_2 D_3$, *b*) $\overline{D}_0 \overline{D}_1 D_2 D_3$, *c*) $\overline{D}_0 \overline{D}_1 D_2 \overline{D}_3$, or *d*) $\overline{D}_0 D_1 \overline{D}_2 D_3$.

Short Exercises

6.72 Explain why c.m.o.s. shift registers are mostly of the SISO type.

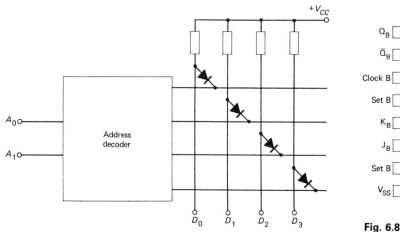

Fig. 6.6

Fig. 6.8

6.73 A diode rom of the type shown in Fig. 6.6 is to produce the Boolean equation

$$F = D_0D_1D_2D_3 + \overline{D}_0\overline{D}_1\overline{D}_2\overline{D}_3 + D_0\overline{D}_1\overline{D}_2D_3 + \overline{D}_0D_1D_2\overline{D}_3$$

Determine the necessary diode connections.

6.74 Draw the block diagram of a 6-bit shift register using J-K flip-flops. Explain how it operates when the number 110010 is loaded *a*) serially and *b*) in parallel. Assume a serial output.

6.75 Explain the function of the shift mode control of a shift register.

6.76 Fig. 6.7 shows the pin connections of the c.m.o.s. 40100 32-stage shift register. State the function of each pin.

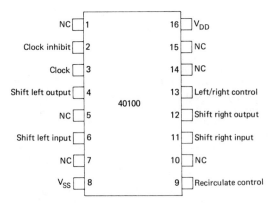

Fig. 6.7

6.77 A prom is listed as having a capacity of 254×4. What does this mean?

6.78 Explain how the programme stored in an eprom can be removed and a new programme installed.

6.79 Fig. 6.8 shows the pin connections of the 4027 dual J-K flip-flop. Show how two of these i.c.s can be connected together to form a 4-bit shift register.

6.80 A 64-bit square memory matrix is addressed by the binary number 110100. In which row and in which column is the wanted location?

6.81 A shift register has 8 stages. If the clock frequency is 2 MHz calculate the time needed to load the register *a*) serially and *b*) in parallel.

6.82 Fig. 6.2 shows the pin connections of the HM 6561 c.m.o.s. static ram. *a*) How many bits can be stored? *b*) How is the memory organized? *c*) What are the functions of the $\overline{CS}$ and $\overline{E}$ pins? *d*) Do the pins in *c*) have to go high or low to be active?

6.83 A rom has 11 address lines and 8 data lines. Calculate *a*) the number of bits stored and *b*) the organization of the memory.

6.84 Fig. 6.9 shows the basic block diagram of a 1024 × 1 ram. Draw a diagram to show how 4 such rams could be connected to give a 1024 × 4 memory.

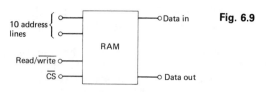

Fig. 6.9

6.85 A rom has 12 address lines. Calculate the number of memory locations.

6.86 The main factors affecting the choice of a ram for a particular application are *a*) the memory size required, *b*) the organization of the memory, and *c*) the access time. Explain the meaning of each of these terms.

6.87 A rom is organized as $8K \times 8$. List the function of the necessary i.c. pins. What is the minimum number of pins required?

Answers to self-test questions

2.1 F **2.2** F **2.3** T **2.4** T **2.5** F **2.6** T
2.7 F **2.8** T **2.9** T **2.10** F **2.11** T **2.12** F
2.13 F **2.14** T **2.15** T **2.16** T **2.17** T
2.18 F **2.19** T **2.20** F **2.21** T **2.22** T
2.23 T **2.24** F **2.25** *c* **2.26** *c* **2.27** *b*
2.28 *a* **2.29** *a* **2.30** *d* **2.31** *d* **2.32** *d*
2.33 *d* **2.34** *c* **2.35** *d* **2.36** *b* **2.37** *a*
2.38 *a* **2.39** Fig A1 **2.40** $\overline{C}(A\overline{B} + \overline{A}B)$
$+ C(AB + \overline{A}\overline{B})$ **2.41** Fig A2 **2.42** $A + B\overline{C}$
2.44 $\overline{A}$ **2.45** *a*) $AD(B + C)$, *b*) $(A + B)(C + D)$
2.46 $A = B = E = F = 1, C = D = G = 0$ **2.47** F
2.48 T **2.49** T **2.50** T **2.51** F **2.52** T
2.53 F **2.54** T **2.55** T **2.56** F **2.57** T
2.58 F **2.59** T **2.60** F **2.61** F **2.62** F
2.63 T **2.64** F **2.65** T **2.66** F **2.67** T
2.68 T **2.69** T **2.70** F **2.71** T **2.72** T
2.73 T **2.74** F **2.75** T **2.76** T **2.77** T
2.78 F **2.79** F **2.80** T **2.81** *b* **2.82** *c*
2.83 *b* **2.84** *d* **2.85** *c* **2.86** *d* **2.87** *b*
2.88 *b* **2.89** *d* **2.90** *c* **2.91** *b* **2.92** *c* or *d*
2.93 $\overline{A} + \overline{B} + C + D + \overline{E} + \overline{F} + G$ **2.94** 0
2.95 $\overline{C}(A\overline{B} + \overline{A}B) + C(AB + \overline{A}\overline{B})$
2.99 $(A + B)(A + C)$ **2.102** 0,1,0 **2.104** 0, AB,
$A(B + \overline{C}), D(\overline{A} + \overline{C})$ **2.108** $\overline{A}BC + A\overline{B} + A\overline{C} + \overline{B}C$
2.112 $AC + \overline{B}$ **2.113** $AB + \overline{A}\overline{C}$ **2.114** $AC(\overline{B} + D)$
$+ \overline{A}\overline{C}(B + \overline{D})$ **2.115** $ABC + AB\overline{D} + \overline{B}C$
2.116 $A\overline{B} + C + D$ **2.117** $BD + \overline{B}\overline{D}$
2.118 $\overline{A}C + \overline{C}D + A\overline{C} + \overline{A}B + \overline{B}D$ **2.119** $C + D$
2.120 $BC + BD$ **2.121** 0 **2.122** AB **2.123** F
2.124 T **2.125** T **2.126** T **2.127** F **2.128** F
2.129 T **2.130** T **2.131** F **2.132** T **2.133** *c*
2.134 *c* **2.135** *b* **2.136** *d* **2.137** Fig A3
2.138 $\overline{A}(\overline{B} + \overline{C})$ **2.139** Fig A4
2.140 $(\overline{A} + \overline{B})(\overline{C} + \overline{D})$ **2.142** Fig A5 **2.143** Fig A6
2.144 Fig A7 **2.145** Fig A8 **2.146** Fig A9
2.147 T **2.148** F **2.149** F **2.150** T **2.151** *a*
2.152 *a* **2.153** *d*
2.154 $X = \overline{A}\overline{B}C + \overline{A}BC, Y = A\overline{B} + \overline{A}B\overline{C} + \overline{A}\overline{B}C,$
$Z = AB$ **2.155** $\overline{A}\overline{B} + \overline{C}$ **2.156** $F = A\overline{B}D + AB\overline{C}\overline{D}$
2.157 $F = C$

3.1 T **3.2** F **3.3** F **3.4** T **3.5** F **3.6** F
3.7 F **3.8** T **3.9** T **3.10** T **3.11** F **3.12** F
3.13 T **3.14** F **3.15** F **3.16** T **3.17** T
3.18 T **3.19** *b* **3.20** *d* **3.21** *a* **3.22** *c*
3.23 *d* **3.24** *a* **3.25** *a* **3.26** *c* **3.27** *b*
3.28 *a* **3.29** *c* **3.30** 20 μA **3.31** 0.2 V, 150 ns,
300 ns **3.33** 7 **3.34** 4.425 V **3.35** 400 mV
3.36 1.7 V, 1.3 V **3.37** T **3.38** T **3.39** F
3.40 T **3.41** F **3.42** T **3.43** F **3.44** F
3.45 T **3.46** T **3.47** F **3.48** T **3.49** F
3.50 F **3.51** T (but new developments may soon
challenge this) **3.52** T **3.53** T **3.54** F

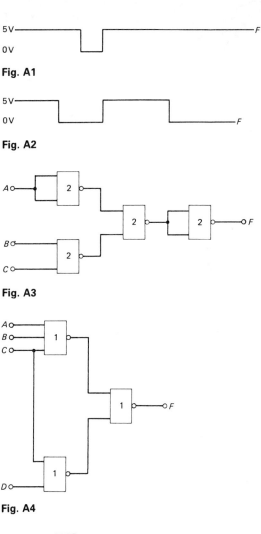

Fig. A1

Fig. A2

Fig. A3

Fig. A4

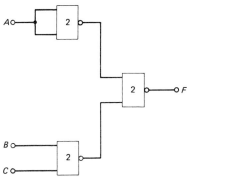

Fig. A5

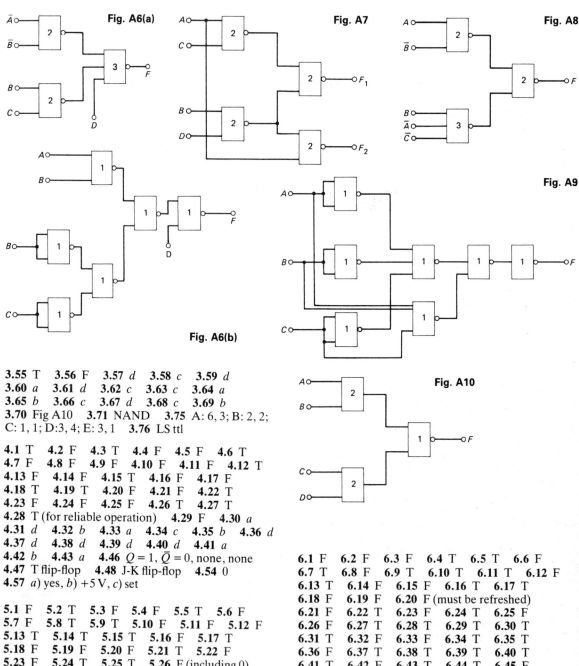

Fig. A6(a)

Fig. A7

Fig. A8

Fig. A6(b)

Fig. A9

Fig. A10

3.55 T **3.56** F **3.57** d **3.58** c **3.59** d
3.60 a **3.61** d **3.62** c **3.63** c **3.64** a
3.65 b **3.66** c **3.67** d **3.68** c **3.69** b
3.70 Fig A10 **3.71** NAND **3.75** A: 6, 3; B: 2, 2;
C: 1, 1; D:3, 4; E: 3, 1 **3.76** LS ttl

4.1 T **4.2** F **4.3** T **4.4** F **4.5** F **4.6** T
4.7 F **4.8** F **4.9** F **4.10** F **4.11** F **4.12** T
4.13 F **4.14** F **4.15** T **4.16** F **4.17** F
4.18 T **4.19** T **4.20** F **4.21** F **4.22** T
4.23 F **4.24** F **4.25** F **4.26** T **4.27** T
4.28 T (for reliable operation) **4.29** F **4.30** a
4.31 d **4.32** b **4.33** a **4.34** c **4.35** b **4.36** d
4.37 d **4.38** d **4.39** d **4.40** d **4.41** a
4.42 b **4.43** a **4.46** $Q = 1, \bar{Q} = 0$, none, none
4.47 T flip-flop **4.48** J-K flip-flop **4.54** 0
4.57 a) yes, b) +5 V, c) set

5.1 F **5.2** T **5.3** F **5.4** F **5.5** T **5.6** F
5.7 F **5.8** T **5.9** T **5.10** F **5.11** F **5.12** F
5.13 T **5.14** T **5.15** T **5.16** F **5.17** T
5.18 F **5.19** F **5.20** F **5.21** T **5.22** F
5.23 F **5.24** T **5.25** T **5.26** F (including 0)
5.27 T **5.28** T **5.29** F **5.30** T **5.31** T
5.32 T **5.33** T **5.34** T **5.35** b **5.36** c
5.37 d **5.38** d **5.39** a **5.40** a **5.41** b
5.42 a **5.43** d **5.44** c **5.45** b **5.48** 3 kHz
5.49 ÷16 **5.50** 0.83 V, 528 ns **5.51** 6
5.52 5 (irregular sequence) **5.54** 16 383
5.62 a) 0–7, b) 7–0 **5.64** 3, 0 **5.65** 6
5.68 500 kHz, 250 kHz, 62.5 kHz

6.1 F **6.2** F **6.3** F **6.4** T **6.5** T **6.6** F
6.7 T **6.8** F **6.9** T **6.10** T **6.11** T **6.12** F
6.13 T **6.14** F **6.15** F **6.16** T **6.17** T
6.18 F **6.19** F **6.20** F (must be refreshed)
6.21 F **6.22** T **6.23** F **6.24** T **6.25** F
6.26 F **6.27** T **6.28** T **6.29** T **6.30** T
6.31 T **6.32** F **6.33** F **6.34** T **6.35** T
6.36 F **6.37** T **6.38** T **6.39** T **6.40** T
6.41 T **6.42** F **6.43** T **6.44** T **6.45** F
6.46 F **6.47** F **6.48** d **6.49** c **6.50** c
6.51 d **6.52** a **6.53** d **6.54** c **6.55** c
6.56 d **6.57** c **6.58** c **6.59** b **6.60** b
6.61 d **6.62** c **6.63** b **6.64** d **6.65** a
6.66 a **6.67** b **6.68** d **6.69** a **6.70** a
6.71 b **6.80** 6, 4 **6.81** a) 4 μs, b) 0.5 μs
6.82 a) 1024, b) 256 × 4, d) low **6.83** a) 16 384,
b) 2048 × 8 **6.85** 4096 **6.87** 24

Index